校企协同软件工程应用型专业“十三五”实训规划系列教材

天津工业大学计算机科学与技术学院
融创软通公司教育培训部
联合编写

Java Web 编程技术

——JSP + Servlet + MVC

陈香凝 / 主　编
张建军　任淑霞 / 副主编

天津大学出版社
TIANJIN UNIVERSITY PRESS

内 容 提 要

本书主要介绍了基于 Java Web 的编程技术，主要包括：Web 技术基础；Servlet 技术模型和 Servlet 容器模型、Servlet API 应用和 Servlet 高级应用；JSP 的各种元素、生命周期、作用域对象等；Web 应用开发中的组件重用技术；EL 表达式语言的使用，JSTL 标准标签库，Web 应用的事件处理与过滤器的应用以及安全性等问题；MVC 开发模式等。同时各个章节都包含小结、经典面试题和跟我上机等内容，让读者的编程技术更加扎实。

本书可作为高等学校计算机及相关专业 Web 编程技术课程的教材，也可供从事基于 Java 的 Web 应用开发技术人员学习参考，同时也可用作 Java Web 开发员国际认证考试的备考材料和培训教材等。

图书在版编目（CIP）数据

Java Web 编程技术：JSP+Servlet+MVC / 陈香凝主编．—天津：天津大学出版社，2019.8（2023.8 重印）

校企协同软件工程应用型专业“十三五”实训规划系列教材

ISBN 978-7-5618-6340-4

Ⅰ．① J…　Ⅱ．①陈…　Ⅲ．① JAVA 语言－程序设计－教材　Ⅳ．① TP312.8

中国版本图书馆 CIP 数据核字 (2019) 第 012018 号

Java Web Biancheng Jishu—JSP+Servlet+MVC

出版发行　天津大学出版社
地　　址　天津市卫津路 92 号天津大学内（邮编：300072)
电　　话　发行部：022-27403647
网　　址　www.tjupress.com.cn
印　　刷　北京盛通商印快线网络科技有限公司
经　　销　全国各地新华书店
开　　本　185mm×260mm
印　　张　17
字　　数　425 千
版　　次　2019 年 8 月第 1 版
印　　次　2023 年 8 月第 4 次
定　　价　46.00 元

前　言

本教材属于校企协同软件工程应用型专业实训系列丛书，是天津工业大学计算机科学与软件学院和融创软通公司的多位教师在近12年的校企协同育人过程中的经验总结并将其不断完善后的成果。

1. 本书编写背景

在多年的教学过程中，作者使用了多本目前市面上已经出版的关于Java Web的教材，这类教材理论讲解准确、充实，但是作者实际了解到的情况是学生听课的时候可以听懂教材里的知识，学完之后却不知道如何应用，独自编写应用功能时无从下手。这几年作者一直潜心研究其中的原因，逐渐发现了问题，并且想了一些方法来解决，还在一些班级进行了实验，效果比较好。现在想把这些想法与其他老师分享，让学生能够学有所用。

2. 阅读本书所需的基础知识

阅读本书需要具有一定的Java基础和HTML基础。具有一定的Java基础意味着读者需要熟悉Java基本语法、熟悉面向对象的概念以及熟悉Java的常用类库。具有一定的HTML基础意味着读者需要掌握HTML文档的基本结构以及常用的标签，掌握CSS和简单的JavaScript语法知识。如果读者有网络相关的知识则更好，对于Web应用的运行机制理解会更深入。

本书由浅入深地构建了知识体系，如果想在Java Web应用中访问数据库、访问属性文件、使用JavaMail发送邮件、上传/下载文件、国际化等功能，这本书还可以作为参考手册。

3. 本书设计思路

本书列出了在Java Web方面由Servlet，JSP，MVC等多个层面的必备知识点，使用了大量的实例来加深读者对于概念的理解，几乎每个知识点都有相应的实例，每一章节都提供了大量的经典面试题和课后习题，帮助读者巩固知识。

本书采用现在各个公司项目开发普遍使用的Maven工具构建Java Web项目，同时各个章节都包含小结、经典面试题和跟我上机等内容，让读者的编程技术更加扎实，提高面试的成功率。

4. 寄语读者

亲爱的读者朋友，感谢您在茫茫书海中找到并选择了本书。您手中的这本教材，不是出自某知名出版社，更不是出自某位名师、大家。它的作者就在您的身边，希望它能够架起你我之间学习、友谊的桥梁，希望它能带您轻松步入妙趣横生的编程世界，希望它会成为您进入 IT 编程行业的奠基石。

Java 技术是无数人经验的积累，希望您通过这本书的学习，能够从一些实例中领悟 Java 开发的精髓，并能够在合适的项目场景下应用它们。有了这本书做参考，将使您在学习过程中得到更多的乐趣。

本书由陈香凝任主编，张建军、任淑霞任副主编。由于时间仓促、学识有限，书中难免有不足和疏漏之处，恳请广大读者将意见和建议通过出版社反馈给我们，以便在后续版本中不断改进和完善。

编者

2018 年 6 月

目录
Contents

Java Web 应用程序开发

本课程目标

- ☐ 理解 Web 应用及开发的基本概念
- ☐ 掌握 Servlet、会话管理、注解配置及 JavaMail 的用法
- ☐ 掌握 JSP 语法、内置对象、标准动作、EL 表达式及 JSTL 标准标签库的使用
- ☐ 掌握 Filter 过滤器、Listener 监听器的用法
- ☐ 掌握文件上传和下载功能的实现
- ☐ 熟练编写基于 MVC（Model II）模式的 Java Web 应用开发
- ☐ 熟练掌握 Maven 项目的配置与应用

第 1 章　Web 应用程序概述

本章要点（学会后请在方框里打钩）：

- □ 了解 Web 相关知识
- □ 了解 Web 应用程序开发相关知识
- □ 了解 Web 服务器相关知识
- □ 掌握 Tomcat 服务器配置过程和使用方法
- □ 掌握 Web 应用程序的目录结构
- □ 掌握 Web 应用程序开发过程

1.1　Web 相关知识

在英语中 Web 即表示网页的意思，它用于表示 Internet 主机上供外界访问的资源。

Internet 上供外界访问的 Web 资源分为以下两种。

（1）静态 Web 资源（如 HTML 页面），指 Web 页面中供人们浏览的数据始终不变。

（2）动态 Web 资源，指 Web 页面中供人们浏览的数据是由程序产生的，不同时间点访问 Web 页面看到的内容各不相同。

专家讲解

（1）静态 Web 资源开发技术：HTML。

（2）常用动态 Web 资源开发技术：JSP/Servlet、ASP.NET、PHP 等。

（3）在 Java 中，动态 Web 资源开发技术统称为 Java Web 技术。

1.2　Web 应用程序

Web 应用程序指供浏览器访问的程序，通常也简称为 Web 应用。例如有 a.html、b.html……多个 Web 资源，这多个 Web 资源用于对外提供服务，此时应把这多个 Web 资源放在一个目录中，以组成一个 Web 应用（或 Web 应用程序）。

一个 Web 应用由多个静态 Web 资源和动态 Web 资源组成，如 html、css、js 文件，JSP 文件、java 程序、支持 jar 包、配置文件等。

Web 应用开发好后，若想供外界访问，需要把 Web 应用所在目录交给 Web 服务器（如 Tomcat）管理，这个过程称为虚拟目录的映射。

Web 应用程序的优点如下。

（1）访问 Web 应用程序更容易。

（2）维护和部署成本低。

1.3　Web 发展史

Web 发展的两个阶段为静态和动态。

1.3.1　静态 Web——HTML

.htm、.html 是网页的后缀，如果现在在一个服务器上直接读取这些内容，那么意味着是把这些网页的内容通过网络服务器展现给用户。整个静态 Web 操作的过程如图 1.1 所示。

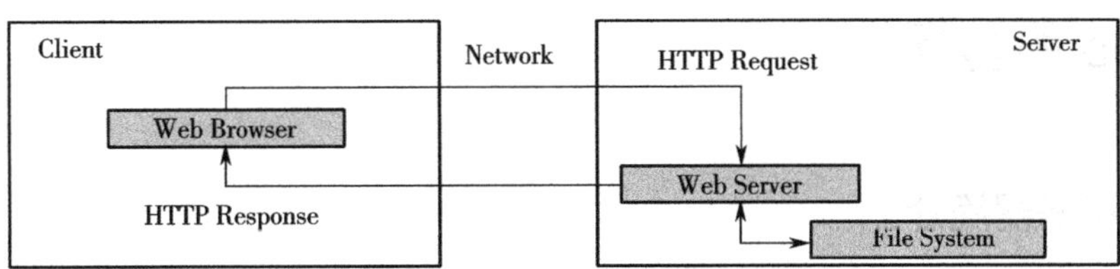

图 1.1 Web 操作过程

在静态 Web 程序中，客户端使用 Web 浏览器（IE、FireFox、Chrome 等）经过网络（Network）连接到服务器上，使用 HTTP 协议发起一个请求（Request），告诉服务器现在需要得到哪个页面，所有的请求交给 Web 服务器，之后 Web 服务器根据用户的需要，从文件系统（存储了所有静态页面的磁盘）取出内容。之后通过 Web 服务器返回给客户端，客户端接收到内容之后经过浏览器渲染解析，得到显示的效果。

静态 Web 存在以下几个缺点。

（1）Web 页面中的内容无法动态更新，所有的用户每时每刻看见的内容和最终效果都是一样的。

为了可以让静态 Web 的显示更加好看，可以加入 JavaScript 以完成一些页面上的显示特效，但是这些特效都是在客户端上借助浏览器展现给用户的，所以在服务器上本身并没有任何变化。

静态 Web 开发以 JavaScript 居多。

（2）静态 Web 无法连接数据库，无法实现和用户的交互（非 Ajax 异步请求方式）。

1.3.2 动态 Web——JSP/Servlet

所谓的动态不是指页面会动，其主要特性是“Web 的页面展示效果因时因人而变”，而且动态 Web 具有交互性，Web 页面的内容可以动态更新。

JSP/Servlet 是 SUN 公司（SUN 公司现在已经被 Oracle 公司收购）主推的 B/S 架构的实现语言，是基于 Java 语言发展起来的，因为 Java 语言足够简单，而且很干净。

JSP/Servlet 技术的性能也是非常好的，不受平台的限制，各个平台基本都可以使用，而且在运行中使用多线程的处理方式，所以性能非常高。

SUN 公司最早推出的 Web 技术是 Servlet 程序，Servlet 程序本身使用的时候有一些问题，程序是采用 Java 代码 +HTML 的方式编写的，即要使用 Java 输出语句，一行一行地输出所有的 HTML 代码。后来，SUN 公司受到了 ASP 的启发，发展出了 JSP（Java Server Pages），JSP 某些代码的编写效果与 ASP 是非常相似的。这样可以很方便地使一些 ASP 程序员转向 JSP 的学习，加大市场的竞争力。

1.4 Web 服务器

1.4.1 Web 服务器简介

（1）Web 服务器是指驻留于因特网上某种类型计算机的程序，是可以向发出请求的浏览器提供文档的程序。当 Web 浏览器（客户端）连接服务器并请求文件时，服务器将处理该请求并将文件反馈到该浏览器上，附带的信息会告诉浏览器如何查看该文件（即文件类型）。Web 服务器请求响应过程如图 1.2 所示。

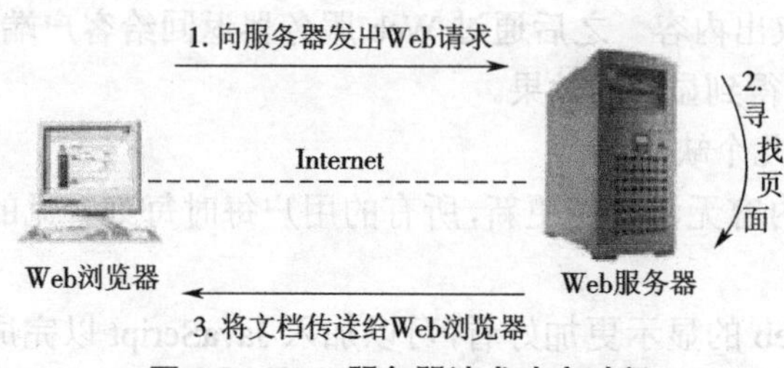

图 1.2 Web 服务器请求响应过程

（2）服务器是一种被动程序：只有当 Internet 上运行在其他计算机中的浏览器发出请求时，服务器才会响应。实现 Web 服务器与客户端交互的过程如图 1.3 所示。

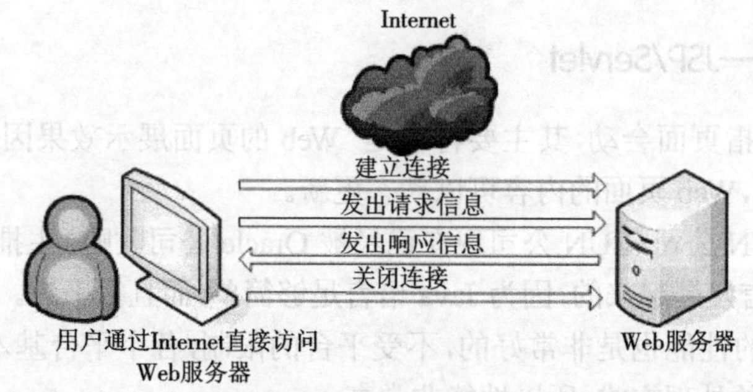

图 1.3 Web 服务器与客户端交互过程

1.4.2 常见的 Web 服务器介绍

1.4.2.1 WebLogic

图 1.4 WebLogic

WebLogic（图 1.4）是美国 Oracle 公司出品的一个 application server，确切地说是一个基于 Java EE 架构的中间件，WebLogic 是用于开发、集成、部署和管理大型分布式 Web 应用、网络应用和数据库应用的 Java 应用服务器，WebLogic 将 Java 的动态功能和 Java Enterprise 标准的安全性引入大型网络应用的开发、集成、部署和管理之中。

WebLogic 是 Oracle 的主要产品之一，系并购 BEA 得来的，是世界上第一个成功商业化的 Java EE 应用服务器，已推出 Oracle WebLogic Server 12cR2（12.2.1）版（下载地址：http://www.oracle.com/technetwork/middleware/weblogic/downloads/index.html）。

1.4.2.2 WebSphere

WebSphere（图 1.5）是 IBM 公司的软件服务平台。WebSphere 提供了可靠、灵活的软件（下载地址：http://www-03.ibm.com/software/products/zh/appserv-was#othertab2）。

图 1.5 WebSphere

1.4.2.3 Tomcat

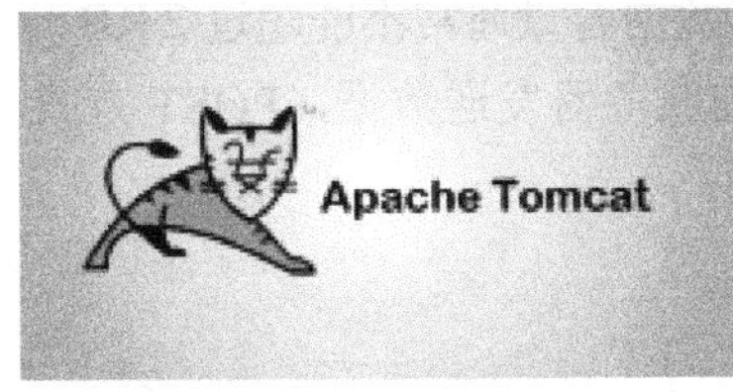

图 1.6 Tomcat

Tomcat（图 1.6）是一个实现了 Java EE 标准的最小的 Web 服务器，是 Apache 软件基金会的 Jakarta 项目中的一个核心项目，由 Apache、Sun 和其他一些公司及个人共同开发而成。因为 Tomcat 技术先进、性能稳定，而且开源免费，因而深受 Java 爱好者的喜爱并得到了部分软件开发商的认可，成为目前比较流行的 Web 应用服务器。学习 Java Web 开发一般都使用 Tomcat 服务器，该服务器支持全部 JSP 以及 Servlet 规范。现在已经推出到了 9.0 版本（下载地址：http://tomcat.apache.org/download-90.cgi）。

专家提醒

学习 Web 开发，需要先安装 Web 服务器，然后在 Web 服务器中部署相应的 Web 资源，供用户使用浏览器访问。

1.5 HTTP 协议

HTTP 是 HyperText Transfer Protocol（超文本传输协议）的简写，它是 TCP/IP 协议的一个应用层协议，用于定义 Web 浏览器与 Web 服务器之间交换数据的过程。客户端连上 Web 服务器后，若想获得 Web 服务器中的某个 Web 资源，需遵守一定的通信格式，HTTP 协议用于定义客户端与 Web 服务器通信的格式。HTTP 协议的版本有 HTTP/1.0 和 HTTP/1.1。

专家讲解

HTTP/1.0 与 HTTP/1.1 的区别如下。

（1）在 HTTP/1.0 协议中，客户端与 Web 服务器建立连接后，只能获得一个 Web 资源。

（2）在 HTTP/1.1 协议中，允许客户端与 Web 服务器建立连接后，在一个连接上获取多个 Web 资源。

1.5.1 HTTP 请求的细节——请求行

请求行中的请求方式有：POST、GET、HEAD、OPTIONS、DELETE、TRACE、PUT。常用的

有：GET、POST。

用户如果没有设置，在默认情况下浏览器向服务器发送的都是 GET 请求，例如在浏览器直接输入地址访问，点超链接访问等都是 GET，用户如果想把请求方式改为 POST，可通过更改表单的提交方式实现。

不管 POST 还是 GET，都用于向服务器请求某个 Web 资源，这两种方式的区别主要表现在数据传递上：如果请求方式为 GET 方式，则可以在请求的 URL 地址后以"?"的形式带上交给服务器的数据，多个数据之间以"&"进行分隔，例如 GET ：/mail/1.html?name=abc&password=xyz HTTP/1.1。

GET 方式的特点：在 URL 地址后附带的参数是有限制的，其数据容量通常不能超过 2 KB。

如果请求方式为 POST 方式，则可以在请求的实体内容中向服务器发送数据，POST 方式的特点是传送的数据量无限制。

专家讲解

GET 与 POST 的区别

提交方式	提交内容的方式	数据量大小	安全性	效率	适用情况
GET	明文，通过 URL 提交数据（数据在 URL 中可以看到）	不超过 2 KB	较低	高	适合提交数据量不大，安全性不高的数据，比如搜索、查询等功能
POST	将用户提交的信息封装在 HTML HEADER 内	数据量大	高	低	适合提交数据量大，安全性高的用户信息，比如注册、修改、上传等功能

1.5.2 HTTP 请求的细节——消息头

HTTP 请求中的常用消息头如下。

（1）Accept：浏览器通过这个头告诉服务器它所支持的数据类型。

（2）Accept-Charset：浏览器通过这个头告诉服务器它支持哪种字符集。

（3）Accept-Encoding：浏览器通过这个头告诉服务器它支持的压缩格式。

（4）Accept-Language：浏览器通过这个头告诉服务器它的语言环境。

（5）Host：浏览器通过这个头告诉服务器它想访问哪台主机。

（6）If-Modified-Since：浏览器通过这个头告诉服务器缓存数据的时间。

（7）Referer：浏览器通过这个头告诉服务器客户机是哪个页面来的防盗链。

（8）Connection：浏览器通过这个头告诉服务器请求完后是断开连接还是保持连接。

1.5.3 HTTP 响应的细节——状态行

状态行格式：HTTP 版本号　状态码　原因叙述 <CRLF>

举例：HTTP/1.1 200 OK

状态码用于表示服务器对请求的处理结果，它是一个三位的十进制数。响应状态码有 5 类，分别为：

（1）100~199 表示成功接收请求，要求客户端继续提交下一次请求才能完成整个处理过程；

（2）200~299 表示成功接收请求并已完成整个处理过程，常用 200；

（3）300~399 表示完成请求，客户需进一步细化请求，例如请求的资源已经移动到一个新地址；

（4）400~499 表示客户端的请求有错误，常用 404；

（5）500~599 表示服务器端出现错误，常用 500。

1.5.4 HTTP 响应细节——常用响应头

HTTP 响应中的常用响应头（消息头）如下。

（1）Location: 服务器通过这个头，告诉浏览器跳到哪里。

（2）Server：服务器通过这个头，告诉浏览器服务器的型号。

（3）Content-Encoding：服务器通过这个头，告诉浏览器，数据的压缩格式。

（4）Content-Length: 服务器通过这个头，告诉浏览器回送数据的长度。

（5）Content-Language: 服务器通过这个头，告诉浏览器语言环境。

（6）Content-Type：服务器通过这个头，告诉浏览器回送数据的类型。

（7）Refresh：服务器通过这个头，告诉浏览器定时刷新。

（8）Content-Disposition: 服务器通过这个头，告诉浏览器以下载方式打数据。

（9）Transfer-Encoding：服务器通过这个头，告诉浏览器数据是以分块方式回送的。

（10）Expires: 默认为 -1，用于指定当前文档应该在什么时候被认为过期。

（11）Cache-Control: no-cache。

（12）Pragma: no-cache。

1.6 配置 Tomcat 服务器

Tomcat 安装配置如下。

（1）下载 Tomcat 8.0.17（下载地址：http://tomcat.apache.org/download-80.cgi）。

（2）添加系统环境变量，我的电脑→属性→高级系统设置→环境变量→新建用户变量。

①新建变量名：TOMCAT_HOME。变量值：D:\Program Files\apache-tomcat-8.0.17（Tomcat 解压到的目录）。

②新建变量名：CATALINA_HOME。变量值：D:\Program Files\apache-tomcat-8.0.17。

③新建变量名：JAVA_HOME。变量值：C:\Program Files\Java\jdk1.8.0_45。

④修改变量名：Path。变量值：%JAVA_HOME%；%JAVA_HOME%\bin；%TOMCAR_HOME%；%TOMCAR_HOME%\bin；% CATALINA_HOME %。

（3）运行 Tomcat 8.0.17，开始→运行→输入 cmd，在命令提示符中输入 startup.bat

【Enter】键后会弹出 Tomcat 命令框，输出启动日志；打开浏览器输入 http://localhost:8080/，如果进入 Tomcat 欢迎界面，那么表示配置成功。

配置成功的界面如图 1.7 所示。

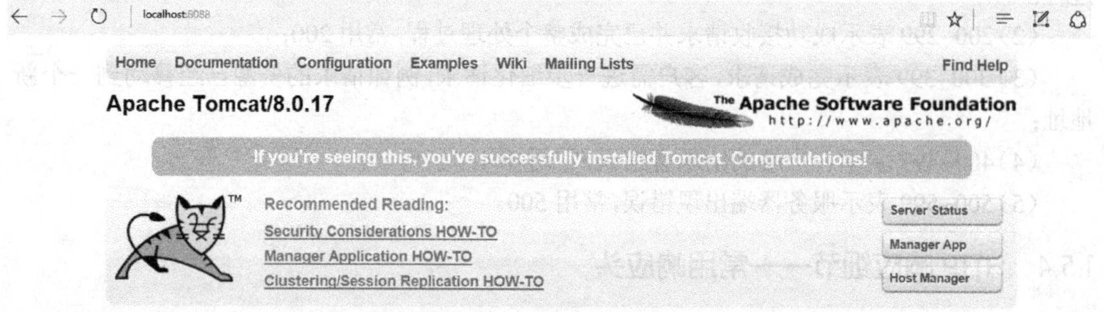

图 1.7　Tomcat 运行结果

（4）配置 Tomcat 用户，找到 Tomcat 目录下的 conf 目录里的 tomcat-users.xml, 加入如下语句：

```
<role rolename="manager-gui"/>
<user username="admin" password="111111" roles="manager-gui"/>
```

（5）重新启动服务器，刷新页面，单击“Manager APP”按钮，输入用户名和密码即可登录。

1.7　Web 应用程序手动开发过程

Web 应用程序手动开发过程如图 1.8 所示。

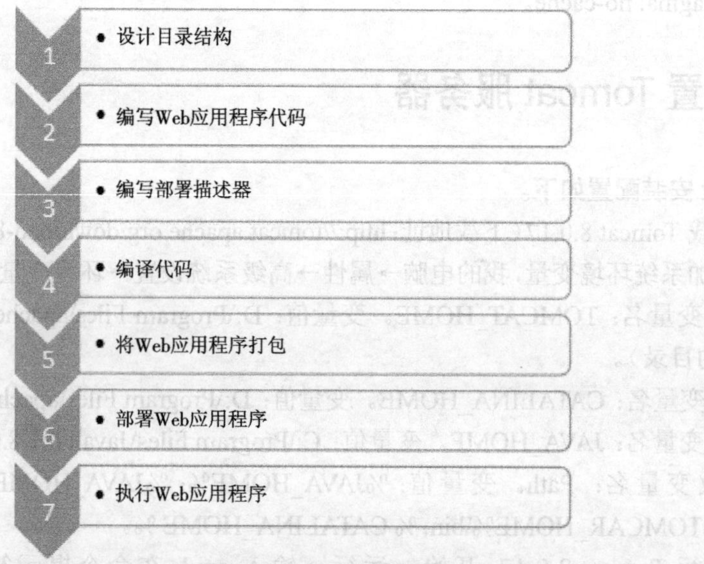

图 1.8　Web 应用程序手动开发过程

1.8 综合案例:Hello World Web 应用程序

1.8.1 建立目录结构

```
WEBROOT : 根目录,一般虚拟目录会直接在此目录中设置
    | --- WEB-INF 整个 Web 中最安全的目录,无法直接访问,若访问,需要在 web.xml
中配置
            |-- classes 保存所有的 *.class 文件的目录,所有的 class 都要放在包中
            |-- lib     存储第三方的 jar 文件的目录
            |-- web.xml Web 的部署描述符文件
    |--- css    存储所有的 *.css 文件的目录
    |--- js     存储所有的 *.js 文件的目录
    |--- jsp    存储所有的 *.jsp 文件的目录
    |--- index.htm, index.jsp 欢迎页面
```

1.8.2 编写 web.xml

```
<?xml version="1.0" encoding="UTF-8"?>
<web-app  version="3.1">
</web-app>
```

1.8.3 编写 Helloworld.jsp

```
<%@ page language="java" import="java.util.*" pageEncoding="UTF-8" %>
<HTML>
<HEAD>
  <TITLE> 我的第一个 Web 应用程序 -HelloWorld!</TITLE>
</HEAD>
<BODY>
<%
  out.println("<h1>Hello World!<br> 世界,你好! </h1>");
%>
</BODY>
</HTML>
```

1.8.4 打包 WAR 文件

控制台命令格式：jar {c t x u f }[v m e 0 M i][-C 目录] 文件名…

其中，{c t x u f} 为必选参数部分。

（1）c：创建一个 jar 包。

（2）t：显示 jar 中的内容列表。

（3）x：解压 jar 包。

（4）u：添加文件到 jar 包中。

（5）f：指定 jar 包的文件名。

[v m e 0 M i] 为可选参数部分。

（1）v：生成详细的报造，并输出至标准设备。

（2）m：指定 manifest.mf 文件。（manifest.mf 文件中可以对 jar 包及其中的内容作一些设置）

（3）0：产生 jar 包时不对其中的内容进行压缩处理。

（4）M：不产生所有文件的清单文件（Manifest.mf）。这个参数会忽略掉 -m 参数的设置。

（5）i：为指定的 jar 文件创建索引文件。

-C 表示转到相应的目录下执行 jar 命令，相当于 cd 到那个目录，然后不带 -C 执行 jar 命令。

```
例如：D:\workspace\protocal>jar -cvf myweb.war *.*
```

1.8.5 启动 Tomcat 发布 WAR 包到服务器

WAR 包发布信息如图 1.9 所示。

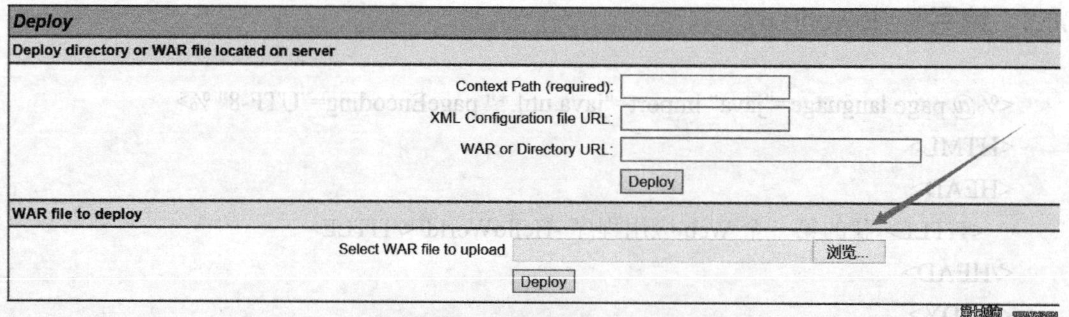

图 1.9 项目部署 WAR 发布信息

1.8.6 访问 Web 项目的 index.jsp 页查看结果

运行结果如图 1.10 所示。

图 1.10 运行结果

小结

本章主要介绍了 Web 应用的概念、HTTP 协议、如何配置 Tomcat 以及手动编写 JavaWeb 应用程序的过程。希望重点掌握如下几点。

(1)掌握静态 Web 和动态 Web 的区别。

(2)熟悉 Web 应用的执行过程,懂得 HTTP 协议的请求细节和响应细节。

(3)熟练配置 Tomcat 和使用 Tomcat。

(4)掌握手动编写 Java Web 应用程序的过程。

经典面试题

1. 什么是动态页面和静态页面?
2. 静态 Web 和动态 Web 的区别与联系是什么?
3. 解释一下什么是 HTTP 协议。
4. 解释 301、404、500、200、304 等 HTTP 状态,它们分别代表什么意思 ?
5. 常用的 Web 服务器软件有哪些?
6. Web 服务器是如何工作的?
7. 描述一下 Tomcat 服务器的配置过程。
8. 手动开发一个 Web 项目的流程是怎样的?
9. 如何配置登录 Tomcat 的用户名和密码?
10. 如何修改 Tomcat 的端口号?

跟我上机

1. 下载 Tomcat 9.0 解压版,手动配置 Tomcat 运行环境,并测试。

2. 手动编写一个简单的 JSP Web 应用:使用记事本工具,完成下图功能(输入一个数,提交表单,显示 *n* 个 Hello,World!)。

注:本章使用的 JSP 技术暂时未讲,请参考第 6 章内容。

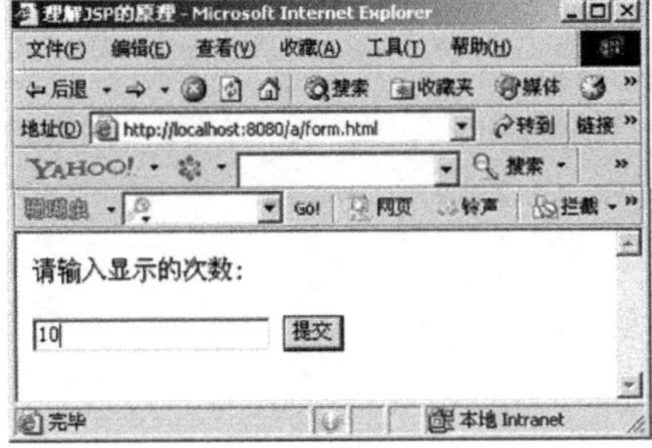

Hello, World!
Hello, World!
Hello, World!
Hello, World!
Hello, World!
Hello, World!
Hello, World!
Hello, World!
Hello, World!
Hello, World!

第 2 章　Servlet 3.0 入门

本章要点（学会后请在方框里打钩）：

□ 了解什么是 Servlet

□ 了解 Servlet 的生命周期

□ 掌握使用 XML 配置 Servlet

□ 掌握使用注解配置 Servlet

□ 掌握使用 Eclipse 工具开发 Servlet

2.1 Servlet 简介

Servlet 最早是由 SUN 公司提供的一门用于开发动态 Web 资源的技术。Servlet 是运行在 Web 服务器上的 Java 程序;它是浏览器(HTTP 客户端)请求和 HTTP 服务器上资源(访问数据库)之间的中间层。

专家提醒

当前较常用版本 Servlet 3.1 是 Java EE7 规范的一部分, Servlet 3.0 以上提供了注解(Annotation)功能,使得不再需要在 web.xml 文件中进行 Servlet 的部署描述,简化了开发流程。

2.2 Servlet 的运行环境

Servlet 容器,如 Tomcat、JBoss、Weblogic、Websphere 等都是 Java Servlet 程序的运行环境。使用 Servlet 可以收集来自网页表单的用户输入,呈现来自数据库或者其他源的记录,还可以动态创建网页。

用户若想实现一个动态 Web 应用,需要完成以下两个步骤。

(1)编写一个 Java 类,实现 Servlet 接口。

(2)把开发好的 Java 类部署到 Web 服务器中。

按照一种约定俗成的称呼习惯,通常也把实现了 Servlet 接口的 Java 程序称为 Servlet。

图 2.1 显示了 Servlet 在 Web 应用程序中的位置。

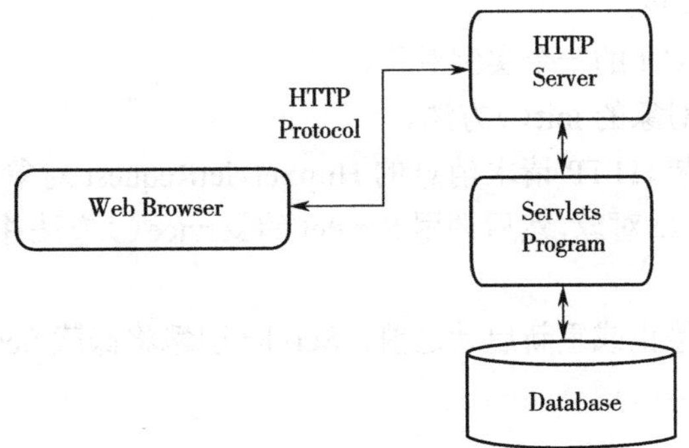

图 2.1 Servlet 在 Web 应用程序中的位置

2.3 Servlet 的生命周期

Servlet 从创建到销毁的过程称为 Servlet 的生命周期。

Servlet 的生命周期可被定义为从创建直到销毁的整个过程。图 2.2 是 Servlet 遵循的过程。

(1)Servlet 通过调用 init()方法进行初始化。

(2)Servlet 通过调用 service()方法来处理客户端的请求。

(3)Servlet 通过调用 destroy()方法终止(结束)。

(4)Servlet 是由 JVM 的垃圾回收器进行垃圾回收的。

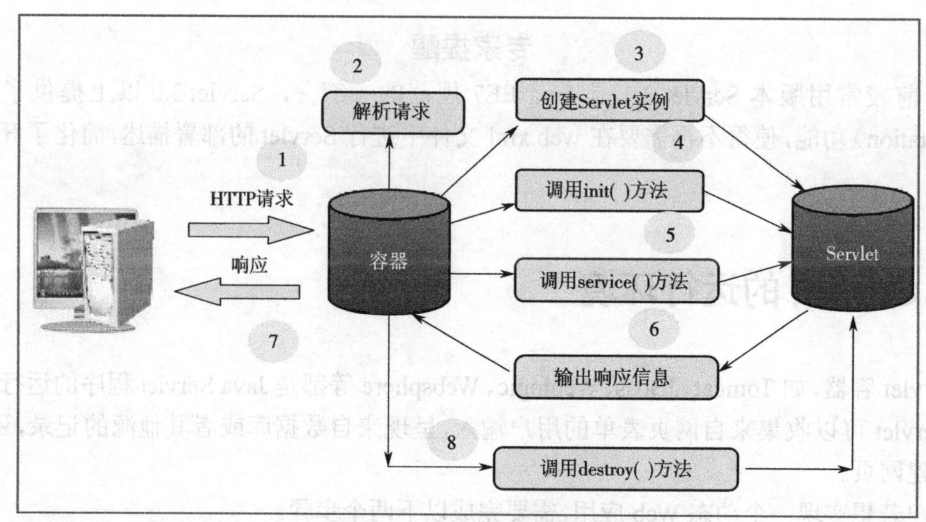

图 2.2 Servlet 的生命周期

Servlet 程序是由 Web 服务器调用的，Web 服务器收到客户端的 Servlet 访问请求后按如下步骤执行。

① Web 服务器首先检查是否已经装载并创建了该 Servlet 的实例对象。如果是，则直接执行第④步，否则，执行第②步。

②装载并创建该 Servlet 的一个实例对象。

③调用 Servlet 实例对象的 init()方法。

④创建一个用于封装 HTTP 请求消息的 HttpServletRequest 对象和一个代表 HTTP 响应消息的 HttpServletResponse 对象，然后调用 Servlet 的 service()方法并将请求和响应对象作为参数传递进去。

⑤ Web 应用程序被停止或重新启动之前，Servlet 引擎将卸载 Servlet，并在卸载之前调用 Servlet 的 destroy()方法。

2.4 简单的 Servlet 结构

Servlet 是服务 HTTP 请求并实现 javax.servlet.Servlet 接口的 Java 类。Web 应用程序开发人员通常编写 Servlet 来扩展 javax.servlet.http.HttpServlet，并实现 Servlet 接口的抽象类专门用于处理 HTTP 请求。

```
package com.isoft.servlet;
import java.io.IOException;
import javax.servlet.ServletConfig;
import javax.servlet.ServletException;
import javax.servlet.ServletRequest;
import javax.servlet.ServletResponse;
import javax.servlet.http.HttpServlet;
public class Hello extends HttpServlet {
    //1. 构造方法
    public Hello() {
            System.out.println("1.servlet 对象被创建了 ");
    }

    // 2.init 方法
    @Override
    public void init(ServletConfig config) throws ServletException {
            System.out.println("2.init 方法被调用 ");
    }

    //3.service 方法
  @Override
    public void service(ServletRequest req, ServletResponse res) throws ServletException,
IOException {
            System.out.println("3.service 方法被调用 ");
    }

    //4.destroy 方法
    @Override
    public void destroy() {
            System.out.println("4.servlet 对象被销毁了 ");
    }
}
```

执行后，效果如图 2.3 所示。

图 2.3 Servlet 生命周期运行结果

以后继续请求时，效果如图 2.4 所示。

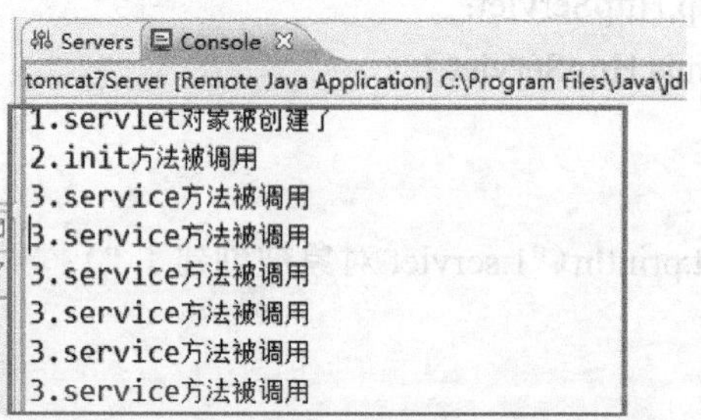

图 2.4 Servlet 生命周期多次请求运行结果

可见，就绪请求时只有 service()方法执行。

2.5 Servlet 的两种配置方式

在 Servlet 2.5 规范之前，Java Web 应用的绝大部分组件都可以通过 web.xml 文件来配置管理，Servlet 3.0 规范以后可通过 Annotation（注解）来配置管理 Web 组件，因此 web.xml 文件变得更加简洁，这也是 Servlet 3.0 的重要简化。

2.5.1 使用 XML 方式配置 Servlet

web.xml 文件中配置 Servlet 如下。

```
<servlet>
  <servlet-name>clientservlet</servlet-name>
  <servlet-class>com.isoft.servlet.ClientServlet</servlet-class>
</servlet>
<servlet-mapping>
  <servlet-name>clientservlet</servlet-name>
  <url-pattern>/clientServlet</url-pattern>
</servlet-mapping>
```

专家提醒

两个 <servlet-name>clientservlet</servlet-name> 名字必须相同。

2.5.2 使用注解方式配置 Servlet

Servlet 3.0 以上的版本提供了注解（Annotation），不再需要在 web.xml 文件中进行 Servlet 的部署描述。

```
@WebServlet("/clientServlet")// 注解配置方式
public class ClientServlet extends HttpServlet {
    private static final long serialVersionUID = 1L;
    protected void doGet(HttpServletRequest request, HttpServletResponse response)
throws ServletException, IOException {
        response.setContentType("text/html; charset=UTF-8");
        PrintWriter out= response.getWriter();
        out.println("<html><body> 您好！ </body></html>");
        out.flush();
        out.close();
    }
    protected void doPost(HttpServletRequest request, HttpServletResponse response)
throws ServletException, IOException {
        doGet(request, response);
    }
}
```

在 Servlet 3.0 中，可以使用 @WebServlet 注解将一个继承于 javax.servlet.http.HttpServlet 的类标注为可以处理用户请求的 Servlet，见表 2.1。

表 2.1 @WebServlet 注解的相关属性

序号	属性名	描　述
1	asyncSupported	声明 Servlet 是否支持异步操作模式
2	description	Servlet 的描述信息
3	displayName	Servlet 的显示名称
3	initParams	Servlet 的初始化参数
5	name	Servlet 的名称
6	urlPatterns	指定一组 Servlet 的 URL 的匹配模式
7	value	等价于 urlPatterns，二者不能共存

Servlet 的访问 URL 是 Servlet 的必选属性，可以选择使用 urlPatterns 或者 value 定义。像上面的 ClientServlet 可以描述成如下形式。

如：@WebServlet（name="Servlet3Demo",value="/clientServlet"）。

如：定义多个 URL 访问，如下：

@WebServlet（name="Servlet3Demo",urlPatterns={"/clientServlet1","/clientServlet2"}）

或 @WebServlet（name="AnnotationServlet",value={"/clientServlet1","/clientServlet2"}）

专家提醒

通过上例对比可以看到开发一个 Servlet 3.1 的程序是非常方便和快捷的。Servlet 3.1 提供了注解之后对于 Servlet 的开发就方便多了，省去了在 web.xml 文件中配置。

2.6 使用 Eclipse 开发 Servlet 应用

（1）打开 Eclipse Java EE 版，新建 Dynamic Web Project 项目，项目名为 Test，如图 2.5 所示。

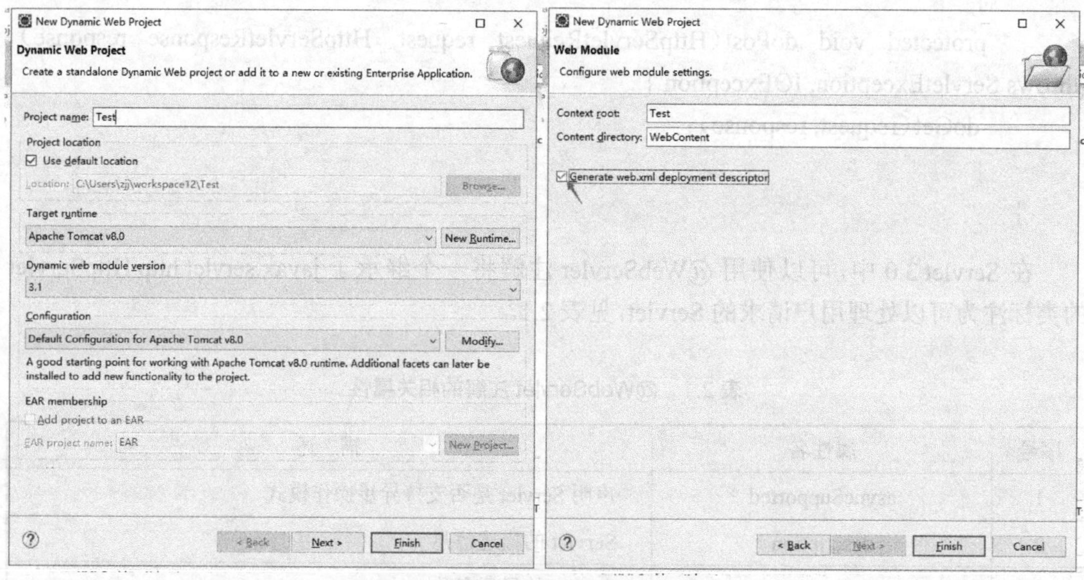

图 2.5 选中 web.xml 配置文件

（2）在 Test 项目的 src 目录下新建一个 package，如 com.iss.servlet，然后新建一个名为 Hello 的 Servlet，如图 2.6 所示。

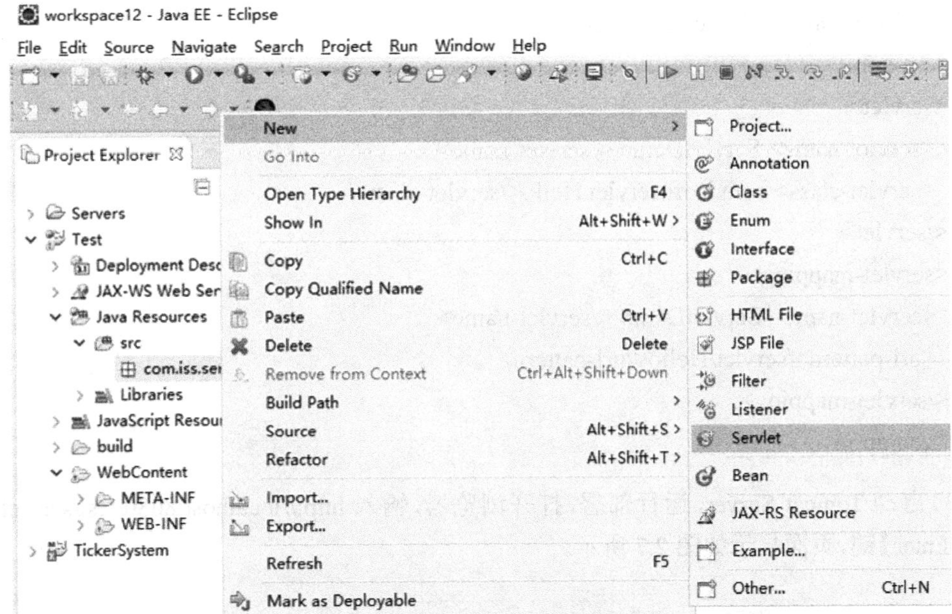

图 2.6 创建 Servlet 类

参考代码如下。

```
//@WebServlet(name="ServletDemo",urlPatterns={"/servlet/Hello "})
public class Hello extends HttpServlet {
   public void doGet(HttpServletRequest request, HttpServletResponse response)
         throws ServletException, IOException {
      response.setContentType("text/html");
      PrintWriter out = response.getWriter();
      out.println("<HTML>");
      out.println("  <HEAD><TITLE>A Servlet</TITLE></HEAD>");
      out.println("  <BODY>");
      out.print("hello world !  This is ");
      out.print(this.getClass());
      out.println(", using the GET method!");
      out.println("</BODY>");
      out.println("</HTML>");
      out.flush();
      out.close();
   }
}
```

与上面的注解方式不同，可以修改 WEB-INF 目录下的 web.xml 文件，代码如下。

```
<?xml version="1.0" encoding="UTF-8"?>
<web-app>
  <servlet>
    <servlet-name> ServletDemo </servlet-name>
    <servlet-class> com.isoft.servlet.Hello</servlet-class>
  </servlet>
  <servlet-mapping>
    <servlet-name> ServletDemo </servlet-name>
    <url-pattern>/servlet/Hello</url-pattern>
  </servlet-mapping>
</web-app>
```

（3）启动 Tomcat Server，运行部署，打开浏览器，输入 http://localhost:8080/test/servlet/Hello，按【Enter】键，页面显示如图 2.7 所示。

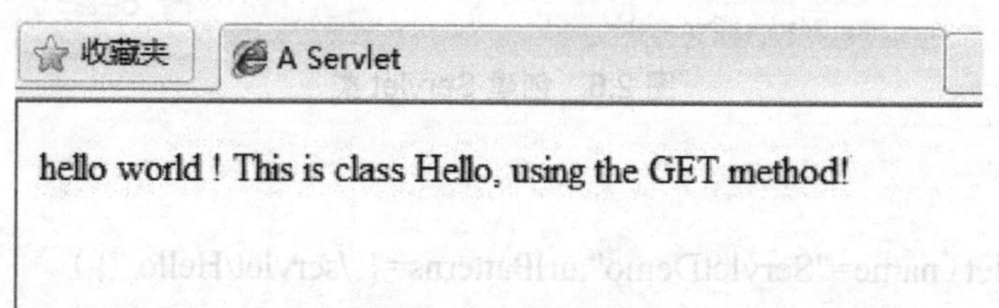

图 2.7　运行结果

2.7　如何开发线程安全的 Servlet

同一个 Servlet 的多个请求，如果该 Servlet 中存在成员变量，可能发生多线程同时访问该资源并都操作它，造成数据的不一致，产生线程安全问题。因此可以通过以下几种方法开发线程安全的 Servlet。

2.7.1　实现 SingleThreadModel 接口

该接口指定了系统对同一个 Servlet 的调用如何处理。如果一个 Servlet 被这个接口指定，那么在这个 Servlet 中的 service 方法将不会有两个线程被同时执行，当然也就不存在线程安全的问题。这种方法需将前面的 ConcurrentTest 类的类头定义更改为如下形式。

```
// SingleThreadModel 已经过期，但是可以用
public class ConcurrentTest extends HttpServlet implements SingleThreadModel {
…
}
```

2.7.2 对共享数据的同步操作

使用 synchronized 关键字能保证一次只有一个线程可以访问被保护的区段，本例中的 Servlet 可以通过同步块操作来保证线程的安全。同步后的代码如下。

```
public class ConcurrentTest extends HttpServlet {
…
PrintWriter output; // 实例变量
public service (HttpServletRequest request, HttpServletResponse response) throws Servlet-
Exception, IOException{
Username = request.getParameter ("username");
Synchronized (this) {
output = response.getWriter ();
try {
Thread. Sleep (5000);
} catch (Interrupted Exception e) {}
output.println (" 用户名 :"+Username+"<BR>");
}
}
}
```

2.7.3 避免使用实例变量

线程安全问题是由实例变量造成的，只要在 Servlet 里面的任何方法都不使用实例变量，那么该 Servlet 就是线程安全的。修正上面的 Servlet 代码，将实例变量改为局部变量实现同样的功能，代码如下。

```
public class ConcurrentTest extends HttpServlet {
public void service (HttpServletRequest request, HttpServletResponse response) throws
ServletException, IOException {
PrintWriter output; // 非实例变量
String username;
response.setContentType ("text/html; charset=utf-8");
…
}
}
```

2.8 综合案例:使用 Servlet 获取表单数据

AcceptUserRegist.java 代码如下。

```
package com.isoft.servlet;
// 导包略
@WebServlet("/acceptUserRegist")
public class AcceptUserRegist extends HttpServlet {
    private static final long serialVersionUID = 1L;
    public String codeToString(String str) // 处理中文字符串的函数
    {
            String s = str;
            try {
                    byte tempB[] = s.getBytes("ISO-8859-1");
                    s = new String(tempB, "utf-8");
                    return s;
            } catch(Exception e) {
                    return s;
            }
    }
    public void init(ServletConfig config) throws ServletException {
            super.init(config);
    }
    public void doPost(HttpServletRequest request, HttpServletResponse response) throws
ServletException, IOException {
            // 设置 mime
            response.setContentType("text/html; charset=utf-8");
            PrintWriter out = response.getWriter();
            out.println("<html><head><title> 新用户注册 </title></head> <body>");
            out.println(" 这是新用户注册所提交的数据 :<br>");
            out.println(" 用户名是 :" + codeToString(request.getParameter("username"))
+ "<br>");
    out.println(" 密 码 是 :" + codeToString(request.getParameter("userpassword")) +
"<br>");
            out.println(" 性 别 是 :" + codeToString(request.getParameter("sex")) +
"<br>");
```

```
            out.println(" 出生年月是 :" + request.getParameter("year") + request.getParameter("month")
                          + request.getParameter("day") + "<br>");
            out.println(" 电子邮箱是 :" + request.getParameter("E-mail") + "<br>");
            out.println(" 家庭住址是 :" + codeToString(request.getParameter("address"))
+ "<br>");
            out.print("</body> </html>");
        }
    }
```

servletform.jsp 代码如下。

```
<%@ page contentType="text/html; charset=utf-8" %>
<script language="javascript">
 function on_submit() // 验证数据的合法性
 {
  if(form1.username.value=="")
  {
   alert(" 用户名不能为空,请输入用户名 !");
   form1.username.focus();
   return false;
  }
  if(form1.userpassword.value=="")
  {
   alert(" 用户密码不能为空,请输入密码 !");
   form1.userpassword.focus();
   return false;
  }
  if(form1.reuserpassword.value=="")
  {
   alert(" 用户确认密码不能为空,请输入密码 !");
   form1.reuserpassword.focus();
   return false;
 }
   if(form1.userpassword.value!=form1.reuserpassword.value)
   {
   alert(" 密码与确认密码不同 ");
```

```
            form1.userpassword.focus();
            return false;
          }
          if(form1.email.value.length!=0)
          {
            for(i=0; i<form1.email.value.length; i++)
            {
              if(form1.email.value.charAt(i)=='@')
              {
                break;
              }
            }
            if(i==form1.email.value.length)
            {
              alert("非法 E-mail 地址!");
              form1.email.focus();
              return false;
            }
          }
          else
          {
            alert("请输入 E-mail!");
            form1.email.focus();
            return false;
          }
        }
      </script>
      <html>
      <head>
      <title>新用户注册</title>
      </head>
      <body>
      <form method="POST" action="acceptUserRegist" name="form1" onsubmit="return on_submit()">
```

```
新用户注册 <br>
用户名(*):<input type="text" name="username" size="20"><br>
密 码(*):<input type="password" name="userpassword" size="20"><br>
再输一次密码(*):<input type="password" name="reuserpassword" size="20"> <br>
性 别 :<input type="radio" value=" 男 " checked name="sex"> 男 <input type="radio" name="sex" value=" 女 "> 女 <br>
出生年月 :<input name="year" size="4" maxlength=4> 年
    <select name="month">
        <option value="1" selected>1</option>
        <option value="2">2</option>
        <option value="3">3</option>
        <option value="4">4</option>
        <option value="5">5</option>
        <option value="6">6</option>
        <option value="7">7</option>
        <option value="8">8</option>
        <option value="9">9</option>
        <option value="10">10</option>
        <option value="11">11</option>
        <option value="12">12</option>
    </select> 月
    <input name="day" size="3" maxlength=4> 日 <br>
电子邮箱(*):<input name="E-mail" maxlength=28><br>
家庭住址 :<input type="text" name="address" size="20"><br>
<input type="submit" value=" 提交 " name="B1"><input type="reset" valu= 全部重写 " name="B2"><br>
</form>
</body>
</html>
```

运行结果如图 2.8 所示。

http://localhost:8081/Chart2_Servelt/servletForm.jsp

新用户注册
用户名(*):融创软通IT学院
密 码(*):●●●●●●
再输一次密码(*):●●●●●●
性别:◉男 ○女
出生年月:2015 年 10 月 30 日
电子邮箱(*):rcrt@91isoft.com
家庭住址:天津市西青区
提交 全部重写

http://localhost:8081/Chart2_Servelt/acceptUserRegist

这是新用户注册所提交的数据:
用户名是:融创软通IT学院
密码是:123456
性别是:男
出生年月是:20151030
电子邮箱是:rcrt@91isoft.com
家庭住址是:天津市西青区

图 2.8　运行结果

小结

本章主要讲解了 Servlet 的生命周期，xml 的配置方式和注解配置方式,主要包括以下知识点。

（1）一个 Servlet 就是 Java 编程语言中的一个类,运行于服务器上,它接受请求并用请求的数据响应客户端。

（2）HTTP 是一个广泛使用的协议,用于在客户端和服务器之间传输数据。

（3）Servlet 使用户可以在服务器上运行 Java 代码和生成动态内容。

（4）HTTP Servlet 发送 HTTP 请求和接受 HTTP 响应。

（5）Servlet 生命周期由三种方法组成,即 init()、service() 和 destroy()。

（6）Servlet API 包含于两个包中,分别为 javax.servlet 和 javax.servlet.http。

（7）Servlet 继承 GenericServlet 类或 HttpServlet 类。

（8）Eclipse 和 Intellij IDEA 都是用于开发基于 Java 的应用程序(如 Servlet、Applet、JavaBean 和 JSP 页面)的工具。

（9）掌握使用 request 获取表单数据。

（10）掌握使用配置文件方式和注解方式配置 Servlet。

经典面试题

1. 什么是 Servlet?

2. 描述一下 Servlet 的生命周期。

3.Servlet 的 init()方法有什么特点?

4.Servlet 是线程安全的吗？

5. 如何使用 XML 配置 Servlet？

6. 如何使用注解配置 Servlet？

7.@WebServlet 的参数都有哪些？

8. 列举 Servlet 中 request 的 10 种以上的方法。

9. 解释一下 Servlet 中的 doGet 和 doPost 方法。

10.Servlet 的父类是什么？

跟我上机

1. 编写一个 Servlet, 显示客户端的 IP 地址。

2. 编写一个含有 init()、service()和 destory()方法的简单 Servlet 程序测试生命周期。

3. 编写一个使用 request 对象获取客户端表单信息的 Servlet。

4. 编写一个 Servlet 显示客户端的区域信息(包括国家和语言)。(提示:使用 request 对象的 getLocales()和 getLocale()方法)

5. 编写一个线程安全的 Servlet,以便显示本 Servlet 被访问的次数。

6. 使用 Servlet 开发一个简单应用程序,用于在用户单击浏览器的刷新按钮时刷新当前系统时间。

7. 使用 Intellij IDEA 工具编写一个 Servlet,实现用户注册功能,其中 Servlet 要接收用户输入的用户名、密码、姓名、性别和年龄,并跳转到 JSP 页面上显示。

8. 编写一个单的 Servlet 输出客户端请求的各种请求信息。(提示:使用 request 对象的 getHeaderNames()方法)

第 3 章　Servlet 请求和响应

本章要点(学会后请在方框里打钩):

- □ 了解 HttpServletRequest 的用法
- □ 使用 HttpServletRequest 解决提交表单乱码问题
- □ 掌握请求和响应的区别
- □ 了解 HttpServletResponse 的用法
- □ 掌握 HttpSession 的应用和原理
- □ 掌握 session 类对象的销毁
- □ 掌握 session 类对象在项目中的应用
- □ 掌握跳转和重定向的区别

3.1 HttpServletRequest 对象介绍

HttpServletRequest 对象代表客户端的请求，当客户端通过 HTTP 协议访问服务器时，HTTP 请求头中的所有信息都封装在这个对象中，通过这个对象提供的方法，可以获得客户端请求的所有信息。

3.1.1 Request 常用方法——获得客户机信息

实例 3.1：通过 Request 对象获取客户端请求信息

```
@WebServlet("/requestDemo1")
public class RequestDemo1 extends HttpServlet {
    public void doGet(HttpServletRequest request, HttpServletResponse response) throws ServletException, IOException {
        String requestUrl = request.getRequestURL().toString(); // 得到请求的 URL 地址
        String requestUri = request.getRequestURI(); // 得到请求的资源
        String queryString = request.getQueryString(); // 得到请求的 URL 地址中附带的参数
        String remoteAddr = request.getRemoteAddr(); // 得到来访者的 IP 地址
        String remoteHost = request.getRemoteHost();
        int remotePort = request.getRemotePort();
        String remoteUser = request.getRemoteUser();
        String method = request.getMethod(); // 得到请求 URL 地址时使用的方法
        String pathInfo = request.getPathInfo();
        String localAddr = request.getLocalAddr(); // 获取 Web 服务器的 IP 地址
        String localName = request.getLocalName(); // 获取 Web 服务器的主机名
        response.setCharacterEncoding("UTF-8"); // 设置将字符以 "UTF-8" 编码输出到客户端浏览器
        // 通过设置响应头控制浏览器以 UTF-8 的编码显示数据，如果不加这句话，那么浏览器显示的将是乱码
        response.setHeader("content-type", "text/html; charset=UTF-8");
        PrintWriter out = response.getWriter();
        out.write("获取到的客户机信息如下：");
        out.write("<hr/>");
        out.write("请求的 URL 地址：" + requestUrl);
        out.write("<br/>");
```

```
            out.write(" 请求的资源：" + requestUri);
            out.write("<br/>");
            out.write(" 请求的 URL 地址中附带的参数：" + queryString);
            out.write("<br/>");
            out.write(" 来访者的 IP 地址：" + remoteAddr);
            out.write("<br/>");
            out.write(" 来访者的主机名：" + remoteHost);
            out.write("<br/>");
            out.write(" 使用的端口号：" + remotePort);
            out.write("<br/>");
            out.write("remoteUser：" + remoteUser);
            out.write("<br/>");
            out.write(" 请求使用的方法：" + method);
            out.write("<br/>");
            out.write("pathInfo：" + pathInfo);
            out.write("<br/>");
            out.write("localAddr：" + localAddr);
            out.write("<br/>");
            out.write("localName：" + localName);
    }
}
```

运行结果如图 3.1 所示。

```
http://localhost:8081/Chart3_Servlet_Demo/requestDemo1

获取到的客户机信息如下：

请求的URL地址：http://localhost:8081/Chart3_Servlet_Demo/requestDemo1
请求的资源：/Chart3_Servlet_Demo/requestDemo1
请求的URL地址中附带的参数：null
来访者的IP地址：0:0:0:0:0:0:0:1
来访者的主机名：0:0:0:0:0:0:0:1
使用的端口号：62095
remoteUser：null
请求使用的方法：GET
pathInfo：null
localAddr：0:0:0:0:0:0:0:1
localName：0:0:0:0:0:0:0:1
```

图 3.1　运行结果

3.1.2 获得客户机请求头信息

实例 3.2：通过 Request 对象获取客户端请求头信息

```
@WebServlet("/requestDemo2")
public class RequestDemo2 extends HttpServlet {
    public void doGet(HttpServletRequest request, HttpServletResponse response) throws
ServletException, IOException {
        response.setCharacterEncoding("UTF-8"); // 设置将字符以 "UTF-8" 编码输
出到客户端浏览器
        // 通过设置响应头控制浏览器以 UTF-8 的编码显示数据
        response.setHeader("content-type", "text/html; charset=UTF-8");
        PrintWriter out = response.getWriter();
        Enumeration<String> reqHeadInfos = request.getHeaderNames(); // 获取所有
的请求头
        out.write(" 获取到的客户端所有的请求头信息如下：");
        out.write("<hr/>");
        while (reqHeadInfos.hasMoreElements()) {
            String headName = (String) reqHeadInfos.nextElement();
            String headValue = request.getHeader(headName); // 根据请求头的名
字获取对应的请求头的值
            out.write(headName + ":" + headValue);
            out.write("<br/>");
        }
        out.write("<br/>");
        out.write(" 获取到的客户端 Accept-Encoding 请求头的值：");
        out.write("<hr/>");
        String value = request.getHeader("Accept-Encoding"); // 获取 Accept-Encoding
请求头对应的值
        out.write(value);
        Enumeration<String> e = request.getHeaders("Accept-Encoding");
        while (e.hasMoreElements()) {
            String string = (String) e.nextElement();
            System.out.println(string);
        }
    }
}
```

运行结果如图 3.2 所示。

http://localhost:8081/Chart3_Servlet_Demo/requestDemo2

获取到的客户端所有的请求头信息如下：

accept:image/gif, image/jpeg, image/pjpeg, application/x-ms-application, application/xaml+xml, application/x-ms-xbap, application/vnd.ms-excel, application/vnd.ms-powerpoint, application/msword, */*
accept-language:en-US,en;q=0.8,zh-Hans-CN;q=0.7,zh-Hans;q=0.5,it-IT;q=0.3,it;q=0.2
cache-control:no-cache
ua-cpu:AMD64
accept-encoding:gzip, deflate
user-agent:Mozilla/5.0 (Windows NT 6.2; Win64; x64; Trident/7.0; rv:11.0) like Gecko
host:localhost:8081
connection:Keep-Alive

获取到的客户端Accept-Encoding请求头的值：

gzip, deflate

图 3.2 运行结果

3.1.3 获得客户机提交请求的数据

获取客户端数据主要采用的方法如下。

（1）getParameter（String）方法（常用）。

（2）getParameterValues（String name）方法（常用）。

（3）getParameterNames（）方法（不常用）。

（4）getParameterMap（）方法（编写框架时常用）。

form1.jsp 代码如下。

```
<%@ page language="java" import="java.util.*" pageEncoding="UTF-8"%>
<!DOCTYPE HTML PUBLIC "-//W3C//DTD HTML 4.01 Transitional//EN">
<html>
<head>
<title>Html 的 Form 表单元素 </title>
</head>
<fieldset style="width: 500px; ">
    <legend>Html 的 Form 表单元素 </legend>
    <form action="requestDemo03" method="post">
            编     号（文本框）： <input type="text" name="userid" value="NO."
                size="2" maxlength="2"><br>
            用户名（文本框）：<input type="text" name="username" value=" 请输入用户名 "><br>
            密     码（密码框）：
            <input type="password" name="userpass" value=" 请输入密码 "><br>
            性     别（单选框）：<input type="radio" name="sex" value=" 男 " checked> 男
```

```
            <input type="radio" name="sex" value=" 女 "> 女 <br>
            部     门(下拉框): <select name="dept">
            <option value=" 技术部 "> 技术部 </option>
            <option value=" 销售部 " selected> 销售部 </option>
            <option value=" 财务部 "> 财务部 </option>
            </select><br>
            兴     趣(复选框): <input type="checkbox" name="inst" value=
" 唱歌 "> 唱歌
            <input type="checkbox" name="inst" value=" 游泳 "> 游泳
            <input type="checkbox" name="inst" value=" 跳舞 "> 跳舞
            <input type="checkbox" name="inst" value=" 编程 " checked> 编程
            <input type="checkbox" name="inst" value=" 上网 "> 上网 <br>
            说     明(文本域):
            <textarea name="note" cols="34" rows="5">
        </textarea>
            <br>
            <input type="hidden" name="hiddenField" value="hiddenvalue" />
            <input type="submit" value=" 提交(提交按钮)">
            <input type="reset" value=" 重置(重置按钮)">
        </form>
    </fieldset>
    </body>
    </html>
```

运行结果如图 3.3 所示。

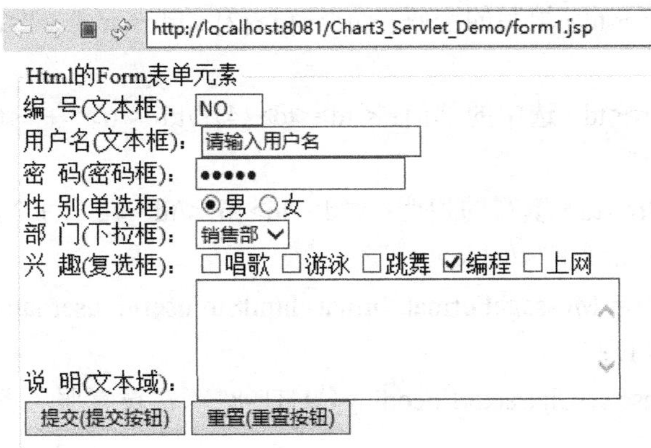

图 3.3　运行结果

在服务器端使用 getParameter 方法和 getParameterValues 方法接收表单参数，代码如下。

```
@WebServlet("/requestDemo3")
public class RequestDemo3 extends HttpServlet {
public void doGet(HttpServletRequest request, HttpServletResponse response) throws ServletException, IOException {
        // 客户端是以 UTF-8 编码提交表单数据的,所以需要设置服务器端以 UTF-8 的编码进行接收,否则对于中文数据就会产生乱码
        request.setCharacterEncoding("UTF-8");
        String userid = request.getParameter("userid");
        String username = request.getParameter("username");
        String userpass = request.getParameter("userpass");
        String sex = request.getParameter("sex");
        String dept = request.getParameter("dept");
        String[] insts = request.getParameterValues("inst");
        String note = request.getParameter("note");
        String hiddenField = request.getParameter("hiddenField");
        String instStr = "";
        for(int i = 0; insts != null && i < insts.length; i++) {
                if(i == insts.length - 1) {
                        instStr += insts[i];
                } else {
                        instStr += insts[i] + ",";
                }
        }
        String htmlStr = "<table>" + "<tr><td> 填写的编号: </td><td>{0}</td></tr>" + "<tr><td> 填写的用户名:</td><td>{1}</td></tr>"
         + "<tr><td> 填写的密码: </td><td>{2}</td></tr>" + "<tr><td> 选中的性别: </td><td>{3}</td></tr>"
         + "<tr><td> 选中的部门: </td><td>{4}</td></tr>" + "<tr><td> 选中的兴趣: </td><td>{5}</td></tr>"
         + "<tr><td> 填写的说明: </td><td>{6}</td></tr>" + "<tr><td> 隐藏域的内容: </td><td>{7}</td></tr>" + "</table>";
        htmlStr = MessageFormat.format(htmlStr, userid, username, userpass, sex, dept, instStr, note, hiddenField);
        response.setCharacterEncoding("UTF-8"); // 设置服务器端以 UTF-8 编码输出数据到客户端
        response.setContentType("text/html; charset=UTF-8"); // 设置客户端浏览器以 UTF-8 编码解析数据
```

```
        response.getWriter().write(htmlStr); // 输出 htmlStr 里面的内容到客户端浏
览器显示
    }
}
```

在服务器端使用 getParameterNames 方法接收表单参数，代码如下。

```
        Enumeration<String> paramNames = request.getParameterNames(); // 获取所
有的参数名
        while (paramNames.hasMoreElements()) {
            String name = paramNames.nextElement(); // 得到参数名
            String value = request.getParameter(name); // 通过参数名获取对应的值
            System.out.println(MessageFormat.format("{0}={1}", name, value));
        }
```

在服务器端使用 getParameterMap 方法接收表单参数，代码如下。

```
  // request 对象封装的参数是以 Map 的形式存储的
        Map<String, String[]> paramMap = request.getParameterMap();
        for (Map.Entry<String, String[]> entry : paramMap.entrySet()) {
            String paramName = entry.getKey();
            String paramValue = "";
            String[] paramValueArr = entry.getValue();
            for (int i = 0; paramValueArr != null && i < paramValueArr.length;
i++) {
                if (i == paramValueArr.length - 1) {
                    paramValue += paramValueArr[i];
                } else {
                    paramValue += paramValueArr[i] + ",";
                }
            }
            System.out.println(MessageFormat.format("{0}={1}",    paramName,
paramValue));
        }
```

3.2 Request 接收表单提交中文参数乱码问题

例如有如下的 form 表单页面：

```
<form action="requestDemo04" method="post">
    用户名:<input type="text" name="userName" />
    <input type="submit" value="post 方式提交表单 ">
    </form>
```

提示:此时在服务器端接收中文参数时会出现中文乱码。

3.2.1 POST 方式提交中文数据乱码产生的原因和解决办法

之所以会产生乱码,是因为服务器和客户端沟通的编码不一致,因此解决的办法是:在客户端和服务器之间设置一个统一的编码,之后就按照此编码进行数据的传输和接收。

使用 request.setCharacterEncoding("UTF-8");设置服务器以 UTF-8 的编码接收数据,就不会产生中文乱码了。

3.2.2 GET 方式提交中文数据乱码产生的原因和解决办法

对于以 GET 方式传输的数据,request 即使设置了以指定的编码接收数据也是无效的,默认的还是使用 ISO8859-1 这个字符编码来接收数据,客户端以 UTF-8 的编码传输数据到服务器端,而服务器端的 request 对象使用的是 ISO8859-1 这个字符编码来接收数据,服务器和客户端沟通的编码不一致因此才会产生中文乱码。解决办法:在接收到数据后,先获取 request 对象以 ISO8859-1 字符编码接收到的原始数据的字节数组,然后通过字节数组以指定的编码构建字符串解决乱码问题。

代码如下。

```
public void doGet(HttpServletRequest request, HttpServletResponse response) throws Serv-
letException, IOException {
    String name = request.getParameter("name");
    name = new String(name.getBytes("ISO8859-1"), "UTF-8");
    System.out.println("name:" + name);
    }
```

3.3 Request 对象实现请求转发

3.3.1 请求转发的基本概念

请求转发,指一个 Web 资源收到客户端请求后,通知服务器去调用另外一个 Web 资源进行处理。

请求转发的应用场景:MVC 设计模式。

在 Servlet 中实现请求转发的两种方式如下。

（1）通过 ServletContext 提供的 getRequestDispatcher（String path）方法，该方法返回一个 RequestDispatcher 对象，调用这个对象的 forward 方法可以实现请求转发。

例如：将请求转发到 test.jsp 页面。

RequestDispatcher rd =this.getServletContext().getRequestDispatcher("/test.jsp");

rd.forward(request, response);

（2）通过 request 对象提供的 getRequestDispatche（String path）方法，该方法返回一个 RequestDispatcher 对象，调用这个对象的 forward 方法可以实现请求转发。

例如：将请求转发到 test.jsp 页面。

request.getRequestDispatcher("/test.jsp").forward(request, response);

request 对象同时也是一个域对象（Map 容器），开发人员通过 request 对象在实现转发时，把数据通过 request 对象带给其他 Web 资源处理。

例如：请求 RequestDemo4 Servlet，RequestDemo4 将请求转发到 test.jsp 页面。

```
@WebServlet("/requestDemo4")
public class RequestDemo4 extends HttpServlet {
    public void doGet(HttpServletRequest request, HttpServletResponse response) throws
ServletException, IOException {
            String data = " 融创软通 IT 学院 ";
            request.setAttribute("data", data);
            request.getRequestDispatcher("/test.jsp").forward(request, response);
    }
}
```

test.jsp 页面代码如下：

```
使用普通方式取出存储在 request 对象中的数据：
<h3 style="color:red;"><%=(String)request.getAttribute("data")%></h3>
使用 EL 表达式取出存储在 request 对象中的数据：
<h3 style="color:red;">${data}</h3>
```

运行结果如图 3.4 所示。

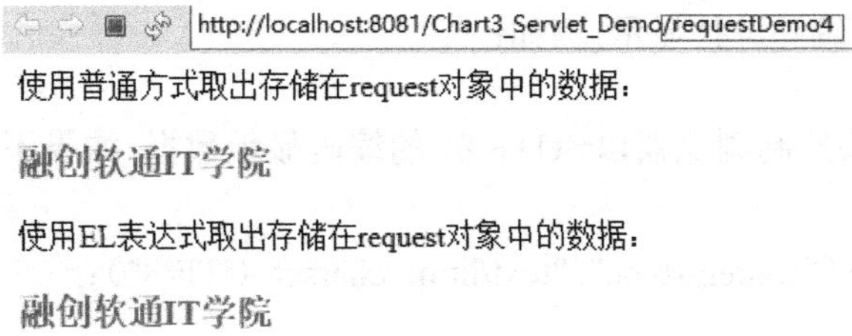

http://localhost:8081/Chart3_Servlet_Demo/requestDemo4

使用普通方式取出存储在request对象中的数据：

融创软通IT学院

使用EL表达式取出存储在request对象中的数据：

融创软通IT学院

图 3.4　运行结果

3.3.2 请求重定向和请求转发的区别

一个 Web 资源收到客户端请求后，通知服务器去调用另外一个 Web 资源进行处理，称为请求转发 /307。

一个 Web 资源收到客户端请求后，通知浏览器去访问另外一个 Web 资源进行处理，称为请求重定向 /302。

3.4 HttpServletResponse 对象介绍

HttpServletResponse 对象代表服务器的响应。这个对象中封装了向客户端发送数据、发送响应头、发送响应状态码的方法。查看 HttpServletResponse 的 API，可以看到相关的方法。

3.4.1 使用 PrintWriter 流向客户端浏览器输出中文数据

使用 PrintWriter 流输出中文应注意的问题：在获取 PrintWriter 输出流之前首先使用“response.setCharacterEncoding（charset）”设置字符以什么样的编码输出到浏览器，如 response.setCharacterEncoding（"UTF-8"）；设置将字符以 "UTF-8" 编码输出到客户端浏览器，然后再使用 response.getWriter（）；获取 PrintWriter 输出流，这两个步骤不能颠倒。

代码如下。

```
@WebServlet("/responseDemo1")
public class ResponseDemo1 extends HttpServlet {
    protected void doGet(HttpServletRequest request, HttpServletResponse response)
                throws ServletException, IOException {
            response.setCharacterEncoding("UTF-8"); // 设置将字符以 "UTF-8" 编码输
出到客户端浏览器
            PrintWriter out = response.getWriter(); // 获取 PrintWriter 输出流
    }
}
```

使用 response.setHeader（"content-type", "text/html; charset= 字符编码 "）；设置响应头，控制浏览器以指定的字符编码进行显示，例如：

```
// 通过设置响应头控制浏览器以 "UTF-8" 的编码显示数据，如果不加这句话，那么浏览
器显示的将是乱码
response.setHeader("content-type", "text/html;charset=UTF-8");
```

除了可以使用 response.setHeader（"content-type", "text/html; charset= 字符编码 "）；设置响应头来控制浏览器以指定的字符编码进行显示这种方式之外，还可以用如下方式来模拟响应头的作用。

```
response.getWriter().write("<meta http-equiv='content-type' content='text/html;charset=UTF-8'/>");
```

实例 3.3：使用 PrintWriter 流向客户端浏览器输出“融创软通 IT 学院：”

```
@WebServlet("/responseDemo2")
public class ResponseDemo2 extends HttpServlet {
    protected void doGet(HttpServletRequest request, HttpServletResponse response)
                throws ServletException, IOException {
            outputOneByPrintWriter(response);
    }
    public void outputOneByPrintWriter(HttpServletResponse response) throws IOExcep-
tion {
            response.setHeader("content-type", "text/html;charset=UTF-8");
            response.setCharacterEncoding("UTF-8");
            PrintWriter out = response.getWriter();// 获取 PrintWriter 输出流
            out.write(" 使用 PrintWriter 流输出数字：融创软通 IT 学院：");
            out.flush();
            out.close();
    }
}
```

运行结果如图 3.5 所示。

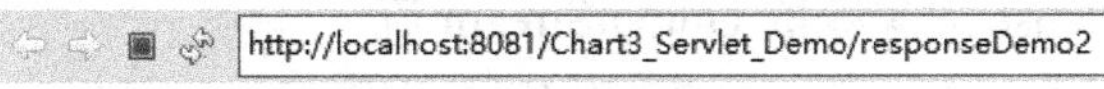

使用PrintWriter流输出数字:融创软通IT学院:

图 3.5 运行结果

3.4.2 使用 Response 实现文件下载

文件下载功能是 Web 开发中经常使用到的功能，使用 HttpServletResponse 对象就可以实现文件的下载。

文件下载功能的实现思路如下。

（1）获取要下载的文件的绝对路径。

（2）获取要下载的文件名。

（3）设置 content-disposition 响应头控制浏览器以下载的形式打开文件。

（4）获取要下载的文件输入流。

（5）创建数据缓冲区。

（6）通过 Response 对象获取 OutputStream 流。

（7）将 FileInputStream 流写入到 buffer 缓冲区。

（8）使用 OutputStream 将缓冲区的数据输出到客户端浏览器。

3.4.3 综合案例:使用 Response 实现文件下载

```
@WebServlet("/responseDemo3")
public class ResponseDemo3 extends HttpServlet {
    public void doGet(HttpServletRequest request, HttpServletResponse response) throws
ServletException, IOException {
        downloadFileByOutputStream(response);// 下载文件,通过 OutputStream 流
    }
    private void downloadFileByOutputStream(HttpServletResponse response) throws File-
NotFoundException, IOException {
        // 1. 获取要下载的文件的绝对路径
        String realPath = this.getServletContext().getRealPath("/download/1.JPG");
        // 2. 获取要下载的文件名
        String fileName = realPath.substring(realPath.lastIndexOf("\\") + 1);
        // 3. 设置 content-disposition 响应头控制浏览器以下载的形式打开文件
        response.setHeader("content-disposition", "attachment;filename=" + fileName);
        // 4. 获取要下载的文件输入流
        InputStream in = new FileInputStream(realPath);
        int len = 0;
        // 5. 创建数据缓冲区
        byte[] buffer = new byte[1024];
        // 6. 通过 Response 对象获取 OutputStream 流
        OutputStream out = response.getOutputStream();
        // 7. 将 FileInputStream 流写入到 buffer 缓冲区
        while ((len = in.read(buffer)) > 0) {
        // 8. 使用 OutputStream 将缓冲区的数据输出到客户端浏览器
            out.write(buffer, 0, len);
        }
        in.close();
    }
}
```

专家提醒

编写文件下载功能时推荐使用 OutputStream 流,避免使用 PrintWriter 流,因为 OutputStream 流是字节流,可以处理任意类型的数据,而 PrintWriter 流是字符流,只能处理字符数据,如果用字符流处理字节数据,会导致数据丢失。这一点千万要注意。

3.5 HttpSession 对象介绍

在 Web 开发中，服务器可以为每个用户浏览器创建一个会话对象（session 对象）。注意：一个浏览器独占一个 session 对象（在默认情况下）。因此，在需要保存用户数据时，服务器程序可以把用户数据写到用户浏览器独占的 session 中，当用户使用浏览器访问其他程序时，其他程序可以从用户的 session 中取出该用户的数据，为用户服务。

专家讲解

session 和 cookie 的主要区别如下。

（1）cookie 是把用户的数据写给用户的浏览器。

（2）session 技术是把用户的数据写到用户独占的 session 中。

（3）session 对象由服务器创建，开发人员可以调用 request 对象的 getSession 方法得到 session 对象。

3.5.1 HttpSession 的实现原理

HttpSession 的实现原理如图 3.6 所示。

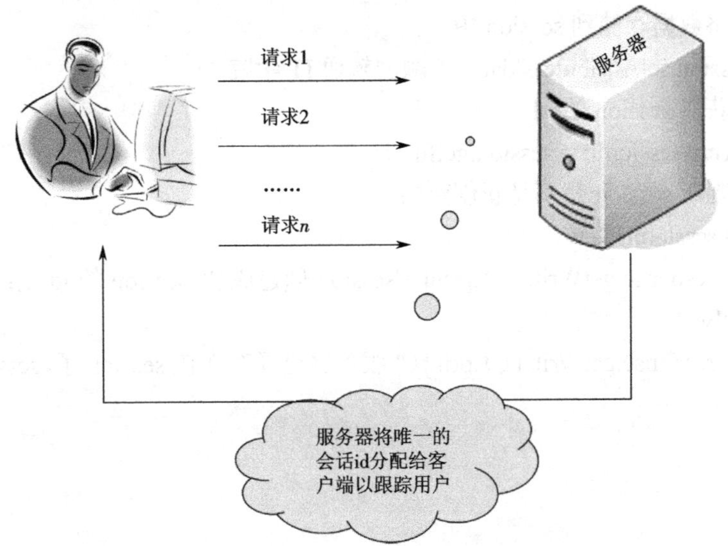

图 3.6 HttpSession 的实现原理

3.5.1.1 服务器是如何实现一个 session 为一个用户浏览器服务的

服务器创建 session 出来后，会把 session 的 id 号以 cookie 的形式回写给客户机，这样，只要客户机的浏览器不关，再去访问服务器时，都会带着 session 的 id 号去，服务器发现客户机浏览器带 session id 过来了，就会使用内存中与之对应的 session 为之服务。

专家讲解

HttpSession 的 getSession()方法用于创建会话。其语法如下。

(1)public HttpSession getSession()。

(2)public HttpSession getSession(boolean value)。

(3)如果没有与当前请求关联的会话,则 getSession( )方法用于创建会话。

(4)如果布尔值为 true 且当前没有与请求关联的会话,则使用 getSession(boolean value)创建会话;如果布尔值为 false,且当前没有与当前请求关联的会话,则返回 null。

实例 3.4:session 的基本使用方法

```
@WebServlet("/sessionDemo1")
public class SessionDemo1 extends HttpServlet {
    protected void doGet(HttpServletRequest request, HttpServletResponse response)
            throws ServletException, IOException {
        response.setCharacterEncoding("UTF-8");
        response.setContentType("text/html;charset=UTF-8");
        //使用 request 对象的 getSession()获取 session,如果 session 不存在则创建一个
        HttpSession session = request.getSession();
        //将数据存储到 session 中
        session.setAttribute("data", "融创软通 IT 学院");
        //获取 session 的 id
        String sessionId = session.getId();
        //判断 session 是不是新创建的
        if(session.isNew()) {
            response.getWriter().print("session 创建成功,session 的 id 是:" + sessionId);
        } else {
            response.getWriter().print("服务器已经存在该 session 了,session 的 id 是:"
+ sessionId);
        }
    }
}
```

第一次访问时,服务器会创建一个新的 session,并且把 session 的 id 以 cookie 的形式发送给客户端浏览器,如图 3.7 所示。

http://localhost:8081/Chart3_Servlet_Demo/sessionDemo1

session创建成功，session的id是：15B2E041C58C11E8D47027AC88FBF412

图 3.7 第一次访问项目运行结果

单击“刷新”按钮，再次请求服务器，此时可以看到浏览器再次请求服务器时，会把存储到 cookie 中的 session 的 id 一起传递到服务器端，如图 3.8 所示。

http://localhost:8081/Chart3_Servlet_Demo/sessionDemo1

服务器已经存在该session了，session的id是：15B2E041C58C11E8D47027AC88FBF412

图 3.8　再次访问项目运行结果

3.5.1.2　使用 session 实现购物车功能

IndexServlet 代码如下。

```
@WebServlet("/indexServlet")
public class IndexServlet extends HttpServlet {
    public void doGet(HttpServletRequest request, HttpServletResponse response) throws
ServletException, IOException {
        response.setContentType("text/html;charset=UTF-8");
        PrintWriter out = response.getWriter();
        // 创建 session
        request.getSession();
        out.write(" 本网站有如下书:<br/>");
        Set<Map.Entry<String, Book>> set = DB.getAll().entrySet();
        for (Map.Entry<String, Book> me : set) {
            Book book = me.getValue();
            String url = request.getContextPath() + "/servlet/BuyServlet?id=" + book.
getId();
          // response. encodeURL(java.lang.String url)用于对表单 action 和超链接的 url
地址进行重写
          url = response.encodeURL(url);// 将超链接的 url 地址进行重写
          out.println(book.getName() + "  <a href='" + url + "'> 购买 </a><br/>");
        }
    }
}
```

运行结果如图 3.9 所示。

http://localhost:8081/Chart3_Servlet_Demo/indexServlet

本网站有如下书：
WebUI交互式网页开发 购买
Java程序设计基础 购买
Oracle11g 数据应用开发 购买
Spring轻量级框架开发 购买
MyBatis持久层框架应用开发 购买

图 3.9　运行结果

BuyServlet 代码如下。

```
@WebServlet("/buyServlet")
public class BuyServlet extends HttpServlet {
    public void doGet(HttpServletRequest request, HttpServletResponse response) throws
ServletException, IOException {
            String id = request.getParameter("id");
            Book book = DB.getAll().get(id); // 得到用户想买的书
            HttpSession session = request.getSession();
            List<Book> list = (List) session.getAttribute("list"); // 得到用户用于保存所
有书的容器
            if(list == null) {
                list = new ArrayList<Book>();
                session.setAttribute("list", list);
            }
            list.add(book);
            // response. encodeRedirectURL(java.lang.String url)用于对 sendRedirect 方
法后的 url 地址进行重写
            String url = response.encodeRedirectURL(request.getContextPath() + "/servlet/
ListCartServlet");
            System.out.println(url);
            response.sendRedirect(url);
    }
}
```

ListCartServlet 代码如下。

```
@WebServlet("/listCartServlet")
public class ListCartServlet extends HttpServlet {
            response.setContentType("text/html;charset=UTF-8");
            PrintWriter out = response.getWriter();
            HttpSession session = request.getSession();
            List<Book> list = (List) session.getAttribute("list");
            if(list == null || list.size() == 0) {
                out.write(" 对不起,您还没有购买任何商品 !!");
                return;
            }
            // 显示用户买过的商品
            out.write(" 您买过如下商品 :<br>");
```

```
            for (Book book : list) {
                out.write(book.getName() + "<br/>");
            }
        }
    }
```

通过查看 IndexServlet 生成的 html 代码可以看到，每一个超链接后面都带上了 session 的 id，如图 3.10 所示。

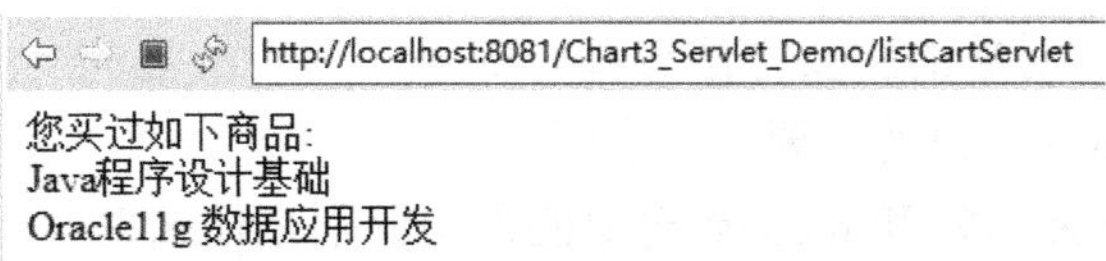

图 3.10 运行结果

3.5.2 session 对象的销毁时机

session 对象默认 30 分钟没有使用，则服务器会自动销毁 session，在 web.xml 文件中也可以手工配置 session 的失效时间，例如：

```
<!-- 设置 session 的有效时间 : 以分钟为单位 -->
<session-config>
<session-timeout>15</session-timeout>
</session-config>
```

当需要在程序中手动设置 session 失效时，可以手工调用 session.invalidate 方法，销毁 session。

```
HttpSession httpSession = request.getSession();
// 手工调用 session.invalidate 方法，销毁 session
session.invalidate();
```

小结

Web 服务器收到客户端的 HTTP 请求时，会针对每一次请求，分别创建一个用于代表请求的 request 对象和代表响应的 response 对象。

request 和 response 对象既然代表请求和响应，如果要获取客户机提交过来的数据，只需要找 request 对象即可，要向客户机输出数据，只需要找 response 对象即可。

javax.servlet.http.HttpSession 接口表示一个会话，可以把一个会话内需要共享的数据保存到 HttpSession 对象中。

HttpServletRequest、HttpSession，ServletContext 都是域对象，一定要掌握它们 3 个之间的区别。

经典面试题

1.HttpServletRequest 是哪个类的对象？

2.HttpServletResponse 是哪个类的对象？

3.HttpServletRequest 对象的主要方法有哪些？

4.HttpServletResponse 对象的主要方法有哪些？

5.HttpServletResponse 对象和 request 对象的作用分别是什么？

6.JavaScript 如何获取 request 对象？

7. 跳转和页面重定向的区别是什么？

8. 通过 HttpServletResponse 对象的什么方法可以设置响应所采用的字符编码？

9. 如何使用 HttpServletResponse 对象给客户端返回 JSON 数据？

10. 怎样从 HttpServletRequest 中得到完整的请求 URL？

跟我上机

1. 编写一个 Servlet，用于显示会话 id 并检索会话为非活动的最大时间段，将此会话设置为 5 分钟后销毁。

2. 编写一个 Servlet 间通信的实例，其中第一个 Servlet 检索远程主机，第二个 Servlet 显示远程主机。

3. 检查会话是否合法：（使用 session）如果登录成功显示欢迎界面；如果不成功显示重新登录界面。

4. 使用隐藏表单域传递商品信息，从而模拟一个购物车。

5.Servlet 检索 html 页上的姓名文本框，并 forward 到第二个 Servlet 上。第二个 Servlet 上显示此人的详细信息。

6. 使用 Sevlet 技术和会话跟踪实现网络聊天的 Web 应用程序。

7. 使用会话跟踪开发一个 Servlet 来销毁会话。每次访问页面都显示当前会话的会话 id，每访问 Servlet 5 次后，必须销毁会话，并显示会话活动的时间。

8. 使用会话跟踪开发一个 Servlet，它将以表格的形式显示会话 id、会话的创建时间、上次访问的时间和上次访问的总页数。如果客户端是第一次访问该站点，则显示一则欢迎且“上次访问的页数”列为 0；如果客户端是第二次访问该站点，则“上次访问的页数”列应增加为 1。

第 4 章　Servlet API 应用

本章要点(学会后请在方框里打钩):

- □ 了解 ServletConfig 应用
- □ 了解 ServletContext 应用
- □ 掌握使用 ServletContext 对象在 Servlet 间传递数据
- □ 掌握使用 ServletContext 读取上下文参数
- □ 掌握使用 Eclipse 和 Intellij IDEA 开发带有验证码功能的用户登录

4.1 ServletConfig 讲解

4.1.1 配置 Servlet 初始化参数

在 Servlet 的配置文件中，可以用一个或多个 <init-param> 标签为 Servlet 配置一些初始化参数。

实例 4.1：配置 Servlet 初始化参数

```
<servlet>
        <servlet-name>ServletConfigDemo1</servlet-name>
        <servlet-class>com.isoft.servlet.ServletConfigDemo</servlet-class>
        <!-- 配置 ServletConfigDemo1 的初始化参数 -->
        <init-param>
                <param-name>uname</param-name>
                <param-value>rcrt</param-value>
        </init-param>
        <init-param>
                <param-name>upwd</param-name>
                <param-value>123456</param-value>
        </init-param>
        <init-param>
                <param-name>charset</param-name>
                <param-value>UTF-8</param-value>
        </init-param>
</servlet>
<servlet-mapping>
        <servlet-name>ServletConfigDemo1</servlet-name>
        <url-pattern>/servletConfigDemo</url-pattern>
</servlet-mapping>
```

4.1.2 通过 ServletConfig 获取 Servlet 的初始化参数

当 Servlet 配置了初始化参数后，Web 容器在创建 Servlet 实例对象时，会自动将这些初始化参数封装到 ServletConfig 对象中，并在调用 Servlet 的 init 方法时，将 ServletConfig 对象传递给 Servlet。进而通过 ServletConfig 对象得到当前 Servlet 的初始化参数信息。

实例 4.2:通过 ServletConfig 获取 Servlet 的初始化参数

```
//@WebServlet("/servletConfigDemo")
public class ServletConfigDemo extends HttpServlet {
    /**
     * 定义 ServletConfig 对象来接收配置的初始化参数
     */
    private ServletConfig config;
    /**
     * 当 Servlet 配置了初始化参数后,Web 容器在创建 Servlet 实例对象时,
     * 会自动将这些初始化参数封装到 ServletConfig 对象中,并在调用 Servlet 的 init 方
法时,
     * 将 ServletConfig 对象传递给 Servlet。进而,程序员通过 ServletConfig 对象就可以
     * 得到当前 Servlet 的初始化参数信息。
     */
    @Override
    public void init(ServletConfig config) throws ServletException {
        this.config = config;
    }
    public void doGet(HttpServletRequest request, HttpServletResponse response) throws
ServletException, IOException {
        // 获取在 web.xml 中配置的初始化参数
        String paramVal = this.config.getInitParameter("uname");// 获取指定的初始化
参数
        response.getWriter().print(paramVal);
        response.getWriter().print("<hr/>");
        // 获取所有的初始化参数
        Enumeration<String> e = config.getInitParameterNames();
        while (e.hasMoreElements()) {
            String name = e.nextElement();
            String value = config.getInitParameter(name);
            response.getWriter().print(name + "=" + value + "<br/>");
        }
    }
    public void doPost(HttpServletRequest request, HttpServletResponse response) throws
ServletException, IOException {
        this.doGet(request, response);
    }
}
```

运行结果如图 4.1 所示。

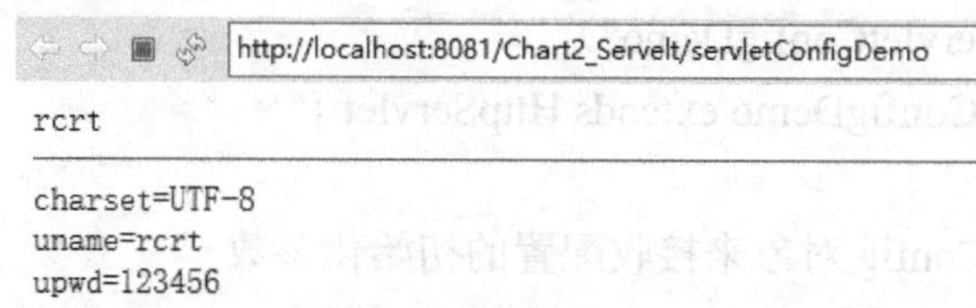

图 4.1 运行结果

4.2 ServletContext 对象

Web 容器在启动时，会为每个 Web 应用程序创建一个对应的 ServletContext 对象，它代表当前的 Web 应用。

ServletConfig 对象中维护了 ServletContext 对象的引用，开发人员在编写 Servlet 时，可以通过 ServletConfig.getServletContext 方法获得 ServletContext 对象。

由于一个 Web 应用中的所有 Servlet 共享同一个 ServletContext 对象，因此 Servlet 对象之间可以通过 ServletContext 对象来实现通信。ServletContext 对象通常也被称为 context 域对象。

4.2.1 ServletContext 的应用

综合案例：多个 Servlet 通过 ServletContext 对象实现数据共享，ServletContextDemo1 和 ServletContextDemo2 通过 ServletContext 对象实现数据共享。

实例 4.3：获得 ServletContext 上下文对象

```
@WebServlet("/servletContextDemo1")
public class ServletContextDemo1 extends HttpServlet {
    public void doGet(HttpServletRequest request, HttpServletResponse response) throws
ServletException, IOException {
            String data = " 融创软通 IT 学院 ";
            /**
             * ServletConfig 对象中维护了 ServletContext 对象的引用，开发人员在编写
Servlet 时，
             * 可以通过 ServletConfig.getServletContext 方法获得 ServletContext 对象。
             */
            ServletContext context = this.getServletConfig().getServletContext();// 获得
ServletContext 对象
            context.setAttribute("data", data); // 将 data 存储到 ServletContext 对象中
    }
```

```
    public void doPost(HttpServletRequest request, HttpServletResponse response) throws
ServletException, IOException {
            doGet(request, response);
    }
}
```

实例 4.4:使用 ServletContext 对象取出数据

```
@WebServlet("/servletContextDemo2")
public class ServletContextDemo2 extends HttpServlet {
    public void doGet(HttpServletRequest request, HttpServletResponse response) throws
ServletException, IOException {
            ServletContext context = this.getServletContext();
            String data = (String) context.getAttribute("data");// 从 ServletContext 对象中
取出数据
            response.getWriter().print("data=" + data);
    }
    public void doPost(HttpServletRequest request, HttpServletResponse response) throws
ServletException, IOException {
            doGet(request, response);
    }
}
```

先运行 ServletContextDemo1,将数据 data 存储到 ServletContext 对象中,然后运行 ServletContextDemo2,可以从 ServletContext 对象中取出数据,实现数据共享。

4.2.2 获取 Web 应用的初始化参数

在 web.xml 文件中使用 <context-param> 标签配置 Web 应用的初始化参数,代码如下。

```
<context-param>
        <param-name>url</param-name>
        <param-value>jdbc:mysql://localhost:3306/test</param-value>
</context-param>
获取 Web 应用的初始化参数,代码如下:
@WebServlet("/servletContextDemo3")
public class ServletContextDemo3 extends HttpServlet {
    public void doGet(HttpServletRequest request, HttpServletResponse response) throws
ServletException, IOException {
```

```
            ServletContext context = this.getServletContext();
            //获取整个 Web 站点的初始化参数
            String contextInitParam = context.getInitParameter("url");
            response.getWriter().print(contextInitParam);
    }
}
```

运行结果如图 4.2 所示。

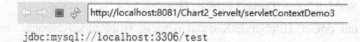

```
http://localhost:8081/Chart2_Servelt/servletContextDemo3
jdbc:mysql://localhost:3306/test
```

图 4.2 运行结果

4.2.3 用 ServletContext 实现请求转发

ServletContextDemo4.java 代码如下。

```
@WebServlet("/servletContextDemo4")
public class ServletContextDemo4 extends HttpServlet {
    public void doGet(HttpServletRequest request, HttpServletResponse response) throws
ServletException, IOException {
        String data = "<h1><font color='red'>www.91isoft.com</font></h1>";
        request.setAttribute("data", data);
        ServletContext context = this.getServletContext();// 获取 ServletContext 对象
        RequestDispatcher rd = context.getRequestDispatcher("/servletContextDemo5");// 获
取请求转发对象(RequestDispatcher)
        rd.forward(request, response);// 调用 forward 方法实现请求转发
    }
}
```

ServletContextDemo5.java 代码如下。

```
@WebServlet("/servletContextDemo5")
public class ServletContextDemo5 extends HttpServlet {
    public void doGet(HttpServletRequest request, HttpServletResponse response) throws
ServletException, IOException {
            response.getOutputStream().print(request.getAttribute("data").toString());
    }
}
```

运行结果如图 4.3 所示。

图 4.3 运行结果

访问的是 ServletContextDemo4，转发到 ServletContextDemo5 并显示了 ServletContextDemo4 的 request 对象的内容，这就是使用 ServletContext 实现了请求转发。

4.2.4 利用 ServletContext 对象读取资源文件

ServletContextDemo6.java 代码如下。

```
@WebServlet("/servletContextDemo6")
public class ServletContextDemo6 extends HttpServlet {
    public void doGet(HttpServletRequest request, HttpServletResponse response) throws ServletException, IOException {
            response.setContentType("text/html;charset=UTF-8");// 目的是控制浏览器用UTF-8 进行解码
            response.setHeader("content-type", "text/html;charset=UTF-8");
            readPropCfgFile2(response);// 读取 src 目录下的 jdbc.properties 配置文件
    }
    private void readPropCfgFile2(HttpServletResponse response) throws IOException {
            InputStream in =
this.getServletContext().getResourceAsStream("/WEB-INF/classes/jdbcConfig.properties");
            Properties prop = new Properties();
            prop.load(in);
            String driver = prop.getProperty("driverClassName");
            String url = prop.getProperty("url");
            String username = prop.getProperty("username");
            String password = prop.getProperty("password");
            response.getWriter().println("读取 src 目录下的 jdbc.properties 配置文件:");
            response.getWriter().println(
                    MessageFormat.format("driver={0},url={1},username={2},password={3}", driver, url, username, password));
    }
}
```

jdbcConfig.properties 代码如下。

```
driverClassName=org.gjt.mm.mysql.Driver
url=jdbc:mysql://localhost:3306/test?useUnicode=true&characterEncoding=utf-8
username=root
password=root
```

运行结果如图 4.4 所示。

http://localhost:8081/Chart2_Servelt/servletContextDemo6

读取src目录下的jdbc.properties配置文件： driver=org.gjt.mm.mysql.Driver,url=jdbc:mysql://localhost:3
useUnicode=true&characterEncoding=utf-8,username=root,password=root

图 4.4 运行结果

4.2.5 使用类装载器读取资源文件

ServletContextDemo7.java 代码如下。

```
public class ServletContextDemo7 extends HttpServlet {
    public void doGet(HttpServletRequest request, HttpServletResponse response) throws
ServletException, IOException {
            response.setHeader("content-type", "text/html;charset=UTF-8");
            test1(response);
    }
    /**
     * 读取类路径下的资源文件
     */
    private void test1(HttpServletResponse response) throws IOException {
            //获取到装载当前类的类装载器
            ClassLoader loader = ServletContextDemo7.class.getClassLoader();
            //用类装载器读取 src 目录下的 db1.properties 配置文件
            InputStream in = loader.getResourceAsStream("jdbcConfig.properties");
            Properties prop = new Properties();
            prop.load(in);
            String driver = prop.getProperty("driverClassName");
            String url = prop.getProperty("url");
            String username = prop.getProperty("username");
            String password = prop.getProperty("password");
            response.getWriter().println("用类装载器读取 src 目录下的 jdbcConfig.
properties 配置文件:");
```

```
            response.getWriter().println(
                        MessageFormat.format("driver={0},url={1},username={2},-
password={3}", driver, url, username, password));
        }
    }
```

运行结果如图 4.5 所示。

http://localhost:8081/Chart2_Servelt/servletContextDemo7

用类装载器读取src目录下的db1.properties配置文件： driver=org.gjt.mm.mysql.Driver,url=jdbc:mysql://localhost:3306/test?useUnicode=true&characterEncoding=utf-8,username=root,password=root

图 4.5 运行结果

4.3 在客户端缓存 Servlet 的输出

对于不经常变化的数据，在 Servlet 中可以为其设置合理的缓存时间值，以避免浏览器频繁向服务器发送请求，提升服务器的性能。例如：

```
    @WebServlet("/servletDemo8")
    public class ServletDemo8 extends HttpServlet {
        public void doGet(HttpServletRequest request, HttpServletResponse response) throws
ServletException, IOException {
            String data = " 融创软通 IT 学院 ";
            /**
             * 设置数据合理的缓存时间值，以避免浏览器频繁向服务器发送请求，提升
服务器的性能。这里是将数据的缓存时间设置为 1 天
             */
            response.setDateHeader("expires", System.currentTimeMillis() + 24 * 3600 *
1000);
            response.getOutputStream().write(data.getBytes());
        }
    }
```

4.4 综合实例：使用 Servlet 生成图片验证码

专家提示

生成验证码图片主要用到了一个 BufferedImage 类。

4.4.1 创建一个 DrawImage Servlet 用来生成验证码图片

```
@WebServlet("/drawImage")
public class DrawImage extends HttpServlet {
    private static final long serialVersionUID = 3038623696184546092L;
    public static final int WIDTH = 120;// 生成的图片的宽度
    public static final int HEIGHT = 30;// 生成的图片的高度
    public void doGet(HttpServletRequest request, HttpServletResponse response) throws
ServletException, IOException {
        this.doPost(request, response);
    }
    public void doPost(HttpServletRequest request, HttpServletResponse response) throws
ServletException, IOException {
        // 1. 在内存中创建一张图片
        BufferedImage bi = new BufferedImage(WIDTH, HEIGHT, BufferedImage.
TYPE_INT_RGB);
        // 2. 得到图片
        Graphics g = bi.getGraphics();
        // 3. 设置图片的背影色
        setBackGround(g);
        // 4. 设置图片的边框
        setBorder(g);
        // 5. 在图片上画干扰线
        drawRandomLine(g);
        String random = drawRandomNum((Graphics2D) g);
        // 6. 根据客户端传递的 createTypeFlag 标识生成验证码图片
        // 7. 将随机数存在 session 中
        request.getSession().setAttribute("checkcode", random);
        // 8. 设置响应头通知浏览器以图片的形式打开
        response.setContentType("image/jpeg");// 等同于 response.setHeader(Content-Type,
image/jpeg);
        // 9. 设置响应头控制浏览器不要缓存
        response.setDateHeader("expries", -1);
        response.setHeader("Cache-Control", "no-cache");
        response.setHeader("Pragma", "no-cache");
        // 10. 将图片写给浏览器
        ImageIO.write(bi, "jpg", response.getOutputStream());
```

```
}
/**
 * 设置图片的背景色
 */
private void setBackGround(Graphics g) {
        g.setColor(Color.WHITE);
        g.fillRect(0, 0, WIDTH, HEIGHT);
}
/**
 * 设置图片的边框
 */
private void setBorder(Graphics g) {
        g.setColor(Color.BLUE);
        g.drawRect(1, 1, WIDTH - 2, HEIGHT - 2);
}
/**
 * 在图片上画随机线条
 */
private void drawRandomLine(Graphics g) {
        g.setColor(Color.GREEN);
        for (int i = 0; i < 5; i++) {
                int x1 = new Random().nextInt(WIDTH);
                int y1 = new Random().nextInt(HEIGHT);
                int x2 = new Random().nextInt(WIDTH);
                int y2 = new Random().nextInt(HEIGHT);
                g.drawLine(x1, y1, x2, y2);
        }
}
private String drawRandomNum(Graphics2D g) {
        g.setColor(Color.RED);
        g.setFont(new Font(" 宋体 ", Font.BOLD, 20));
        String   baseNumLetter   =   "0123456789ABCDEFGHJKLMNOPQRSTU-
VWXYZ";
        // 截取数字和字母的组合
        return createRandomChar(g, baseNumLetter);
}
private String createRandomChar(Graphics2D g, String baseChar) {
```

```
            StringBuffer sb = new StringBuffer();
            int x = 5;
            String ch = "";
            //控制字数
            for (int i = 0; i < 4; i++) {
                //设置字体旋转角度
                int degree = new Random().nextInt() % 30;
                ch = baseChar.charAt(new Random().nextInt(baseChar.length()))
+ "";
                sb.append(ch);
                //正向角度
                g.rotate(degree * Math.PI / 180, x, 20);
                g.drawString(ch, x, 20);
                //反向角度
                g.rotate(-degree * Math.PI / 180, x, 20);
                x += 30;
            }
            return sb.toString();
        }
    }
```

运行结果如图 4.6 所示。

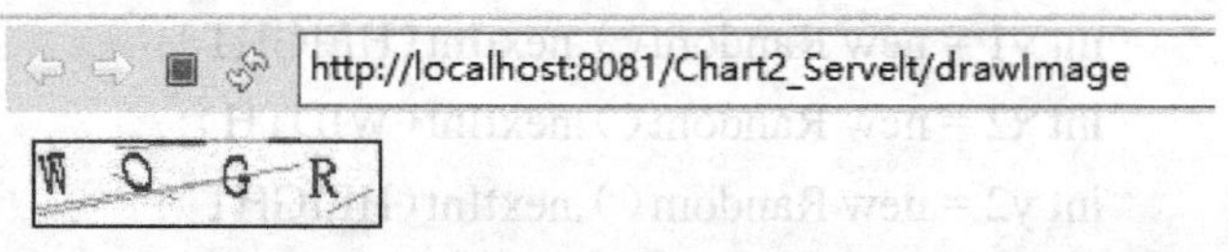

图 4.6 运行结果

4.4.2 在 Form 表单中使用验证码图片

```
<%@ page language="java" import="java.util.*" pageEncoding="UTF-8"%>
<!DOCTYPE HTML PUBLIC "-//W3C//DTD HTML 4.01 Transitional//EN">
<html>
<head>
<title> 在 Form 表单中使用验证码 </title>
<script type="text/javascript">
    //刷新验证码
    function changeImg() {
```

```
            document.getElementById("validateCodeImg").src = "drawImage?id="+
    Math.random();
        }
    </script>
    </head>
    <body>
        <form action="checkServlet" method="post">
        验证码：<input type="text" name="validateCode" /> <img alt="验证码看不清，换
一张"
            src="drawImage" id="validateCodeImg" onclick="changeImg()"> <a
            href="javascript:void(0)" onclick="changeImg()"> 看不清，换一张 </a> <br
/> <input
                type="submit" value=" 提交 ">
        </form>
    </body>
    </html>
```

运行结果如图 4.7 所示。

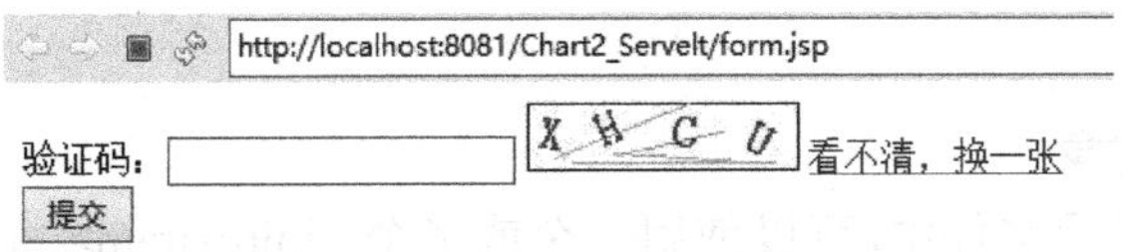

图 4.7　运行结果

4.4.3　服务器端对 form 表单提交上来的验证码处理

```
    @WebServlet("/checkServlet")
    public class CheckServlet extends HttpServlet {
        public void doGet(HttpServletRequest request, HttpServletResponse response) throws
ServletException, IOException {
            String clientCheckcode = request.getParameter("validateCode");// 接收客户端
浏览器提交上来的验证码
            String serverCheckcode = (String) request.getSession().getAttribute("check-
code");// 从服务器端的 session 中取出验证码
            if (clientCheckcode.equals(serverCheckcode)) {// 将客户端验证码和服务器
端验证比较，如果相等，则表示验证通过
                System.out.println(" 验证码验证通过！");
            } else {
```

```
                System.out.println(" 验证码验证失败！");
            }
        }
        public void doPost(HttpServletRequest request, HttpServletResponse response) throws
ServletException, IOException {
                doGet(request, response);
        }
}
```

运行结果如图 4.8 所示。

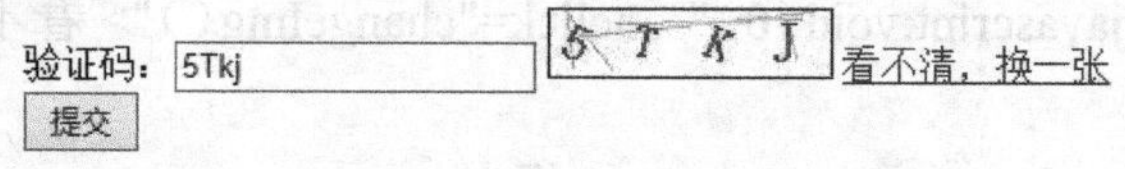

验证码验证通过！

图 4.8　运行结果

小结

1. ServletConfig 对象

（1）在 Servlet 的配置文件中，可以使用一个或多个 <init-param> 标签为 Servlet 配置一些初始化参数。

（2）当 Servlet 配置了初始化参数后，Web 容器在创建 Servlet 实例对象时，会自动将这些初始化参数封装到 ServletConfig 对象传递给 Servlet。进而，程序员通过 ServletConfig 对象就可以得到当前 Servlet 的初始化参数信息。

2. ServletContext 对象

（1）ServletContext 中的属性的生命周期从创建开始，到服务器关闭时结束。

（2）因为存在 ServletContext 中的数据会长时间地保存在服务器中，会占用内存，因此建议不要向 ServeltContext 中添加过大的数据。

经典面试题

1.ServletConfig 在项目中有什么作用？

2. 每个 Servlet 均拥有独立的 ServletConfig 对象吗？

3.Java 中 getServletConfig().getInitParameter(); 的作用是什么？

4.ServletContext 的作用是什么？

5.JSP 中 application 与 ServletContext 有什么区别?
6.ServletContext 和 ServletConfig 有什么区别?
7. 简述 ServletContext 对象保存值的作用范围。
8. 在 web.xml 中,<context-param> 标签的作用是什么?
9.Servlet 中如何获取 ServletContext 对象 ?
10. 解释 Java 的 ServletContext 中的 setAttribute 和 getAttribute 方法。

跟我上机

使用 Servlet 完成在线 QQ 聊天室功能,参考界面如下图所示(提示:页面采用 JSP 页面即可,请参考第 6 章以后的内容或在老师的指导下完成)。

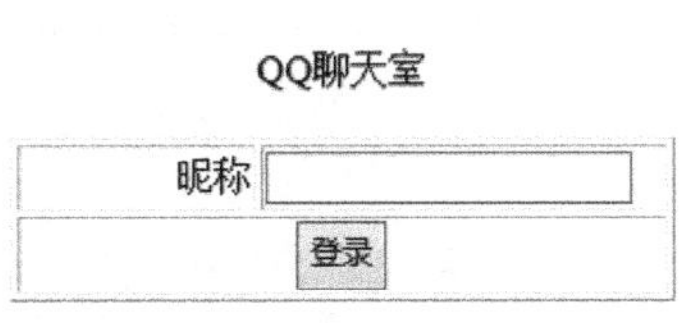

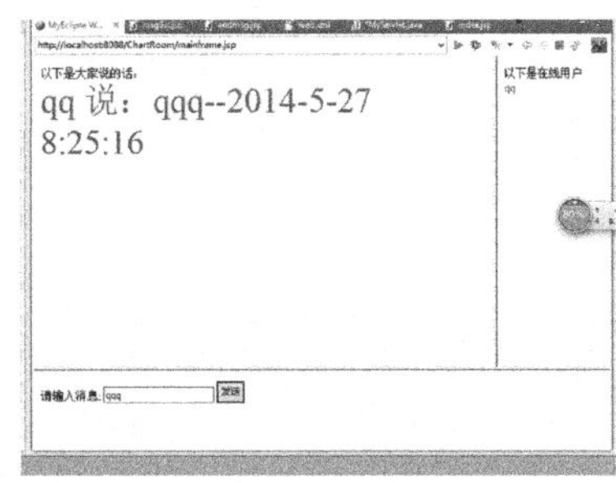

第 5 章　Servlet 高级应用

本章要点(学会后请在方框里打钩):

- □ 掌握 Servlet 中如何接受多个用户请求
- □ 掌握 Servlet 中使用反射机制接受多个用户请求
- □ 掌握 cookie 的用法
- □ 掌握使用 JavaMail 进行邮件的发送和接收

5.1 Servlet 中可以有多个处理请求的方法

一个请求写一个 Servlet 太过复杂和麻烦，我们希望在一个 Servlet 中可以有多个处理请求的方法。因此在客户端发送请求时，必须给出一个参数，用来说明要调用的方法。

如请求路径：http://localhost:8080/baseservlet/base?method=addUser。

专家讲解

当我们访问 Servlet 时，发生了哪些操作？

首先是通过 <url-pattern> 找到 <servlet-name>，通过 <serlvet-name> 最终找到 <servlet-class>，也就是类名，再通过反射机制得到 Serlvet 对象。然后再由 tomcat 调用 init()、service()、destroy()方法，这一过程是 Servlet 的生命周期。在 HttpServlet 里有个 service()的重载方法，ServletRequest、ServletResponse 经过 service(ServletRequest req,ServletResponse resp)方法转换成 HTTP 格式。之后在 service(HttpServletRequest req,HttpServletResponse resp)中调用 doGet()或者 doPost()方法。

实例 5.1：通过请求参数获取要调用的方法

```
@WebServlet("/baseServlet")
public class BaseServlet extends HttpServlet {
    @Override
    protected void service(HttpServletRequest req, HttpServletResponse resp) throws ServletException, IOException {
            String method = req.getParameter("method");// 获取要调用的方法，其中的value="method" 是自己约定的
            if("addUser".equals(method)) {
                    addUser(req, resp);
            }
            if("deleteUser".equals(method)) {
                    deleteUser();
            }
    }
    private void deleteUser() {
            System.out.println(" 删除用户方法被调用 ");
    }
    private void addUser(HttpServletRequest req, HttpServletResponse resp) {
            System.out.println(" 添加用户方法被调用 ");
    }
```

```
}
```

很显然，上面的代码不是我们想要的。

实例 5.2：使用反射机制接受多个用户请求

```
@WebServlet("/baseServlet")
public class BaseServlet extends HttpServlet {
    @Override
    protected void service(HttpServletRequest req, HttpServletResponse resp) throws Serv-
letException, IOException {
            String name = req.getParameter("method");// 获取方法名
            if(name == null || name.isEmpty()) {
                    throw new RuntimeException(" 没有传递 method 参数，请给出你想
调用的方法 ");
            }
            Class c = this.getClass();// 获得当前类的 Class 对象
            Method method = null;
            try {
                    // 获得 Method 对象
                    method = c.getMethod(name, HttpServletRequest.class,
  HttpServletResponse.class);
            } catch (Exception e) {
                    throw new RuntimeException(" 没有找到 " + name + " 方法，请检查
该方法是否存在 ");
            }
            try {
                    method.invoke(this, req, resp);// 反射调用方法
            } catch (Exception e) {
                    System.out.println(" 你调用的方法 " + name + ", 内部发生了异常 ");
                    throw new RuntimeException(e);
            }
    }
}
```

在项目中，用一个 Servlet 继承该 BaseServlet 可以实现多个请求处理。

5.2 使用 cookie 进行会话管理

会话可以简单理解为：用户开一个浏览器，单击多个超链接，访问服务器多个 Web 资源，然后关闭浏览器，整个过程称为一个会话。

有状态会话：一个同学来过教室，下次再来教室时，我们会知道这个同学曾经来过，这称为有状态会话。

cookie 对象常见方法见表 5.1。

表 5.1 cookie 类的主要方法

序 号	方 法	描 述
1	cookie(String name, String value)	实例化 cookie 对象，传入 cookie 名称和值
2	public String getName()	取得 cookie 的名字
3	public String getValue()	取得 cookie 的值
4	public void setValue(String newValue)	设置 cookie 的值
5	public void setMaxAge(int expiry)	设置 cookie 的最长保存时间
6	public int getMaxAge()	获取 cookie 的有效期
7	public void setPath(String uri)	设置 cookie 的有效路径
8	public String getPath()	获取 cookie 的有效路径
9	public void setDomain(String pattern)	设置 cookie 的有效域
10	public String getDomain()	获取 cookie 的有效域

每个用户在使用浏览器与服务器进行会话的过程中，不可避免各自会产生一些数据，程序要想办法为每个用户保存这些数据。

5.2.1 保存会话数据的两种技术

1. cookie

cookie 是客户端技术，程序把每个用户的数据以 cookie 的形式写给用户各自的浏览器。当用户使用浏览器再去访问服务器中的 Web 资源时，就会带着各自的数据。这样，Web 资源处理的就是用户各自的数据了。

2. session

session 是服务器端技术，利用这个技术，服务器在运行时可以为每一个用户的浏览器创建一个其独享的 session 对象，由于 session 为用户浏览器独享，所以用户在访问服务器的 Web 资源时，可以把各自的数据存储在各自的 session 中，当用户再去访问服务器中的其他 Web 资源时，其他 Web 资源再从用户各自的 session 中取出数据为用户服务。

5.2.2 Java 提供的操作 cookie 的 API

Java 中的 javax.servlet.http.Cookie 类用于创建一个 cookie。

response 接口中也定义了一个 addCookie 方法，它用于在其响应头中增加一个相应的 Set-Cookie 头字段。同样，request 接口中也定义了一个 getCookies 方法，它用于获取客户端提交的 cookie。

5.2.3 综合案例：使用 cookie 记录用户上一次访问的时间

```
@WebServlet("/cookieDemo1")
public class cookieDemo1 extends HttpServlet {
    public void doGet(HttpServletRequest request, HttpServletResponse response) throws
ServletException, IOException {
            // 设置服务器端以 UTF-8 编码进行输出
            response.setCharacterEncoding("UTF-8");
            // 设置浏览器以 UTF-8 编码进行接收，解决中文乱码问题
            response.setContentType("text/html;charset=UTF-8");
            PrintWriter out = response.getWriter();
            // 获取浏览器访问服务器时传递过来的 cookie 数组
            Cookie[] cookies = request.getCookies();
            // 如果用户是第一次访问，那么得到的 cookies 将是 null
            if (cookies != null) {
                out.write(" 您上次访问的时间是：");
                for (int i = 0; i < cookies.length; i++) {
                    Cookie cookie = cookies[i];
                    if (cookie.getName().equals("lastAccessTime")) {
                        Long lastAccessTime = Long.parseLong(cookie.get-
Value());

                        Date date = new Date(lastAccessTime);
                        out.write(date.toLocaleString());
                    }
                }
            } else {
                out.write(" 这是您第一次访问本站！");
            }
            // 用户访问过之后重新设置用户的访问时间，存储到 cookie 中，然后发送到
客户端浏览器
```

```
                Cookie cookie = new Cookie("lastAccessTime", System.currentTimeMillis() +
"");// 创建一个 cookie，cookie 的名字是 lastAccessTime
                // 将 cookie 对象添加到 response 对象中，这样服务器在输出 response 对象中
的内容时就会把 cookie 也输出到客户端浏览器
                response.addCookie(cookie);
        }
        public void doPost(HttpServletRequest request, HttpServletResponse response) throws
ServletException, IOException {
                doGet(request, response);
        }
    }
```

第一次访问时这个 Servlet 时，效果如图 5.1 所示。

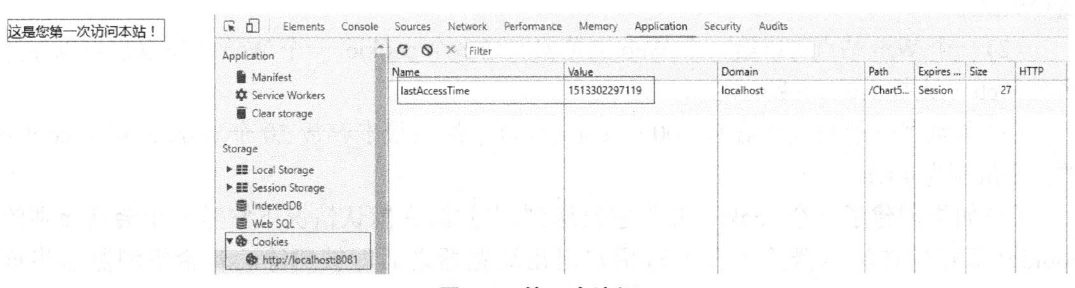

图 5.1　第一次访问

单击浏览器的“刷新”按钮，进行第二次访问，此时服务器就可以通过 cookie 获取浏览器上一次访问的时间了，效果如图 5.2 所示。

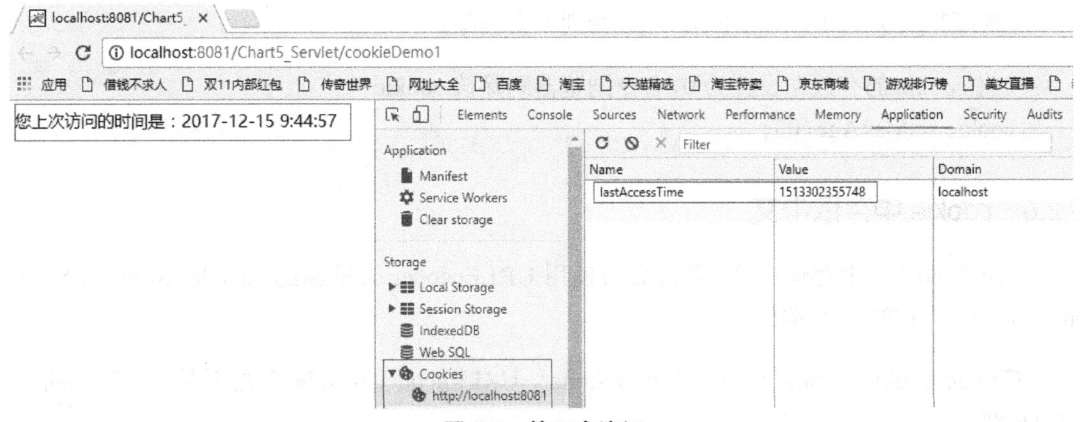

图 5.2　第二次访问

在上面的例子中，程序代码中并没有使用 setMaxAge 方法设置 cookie 的有效期，所以当关闭浏览器之后，cookie 就失效了，要想在关闭了浏览器之后，cookie 依然有效，那么在创建 cookie 时，就要为 cookie 设置一个有效期。如下所示：

```
        Cookie cookie = new Cookie("lastAccessTime", System.currentTimeMillis() +
"");// 创建一个 cookie，cookie 的名字是 lastAccessTime
        // 设置 cookie 的有效期为 1 天
        cookie.setMaxAge(24 * 60 * 60);
        // 将 cookie 对象添加到 response 对象中，这样服务器在输出 response 对象中
的内容时就会把 cookie 也输出到客户端浏览器
        response.addCookie(cookie);
```

这样即使关闭了浏览器，下次再访问时，也依然可以通过 cookie 获取用户上一次访问的时间。

5.2.4 cookie 注意细节

（1）一个 cookie 只能标识一种信息，它至少含有一个标识该信息的名称（NAME）和设置值（VALUE）。

（2）一个 Web 站点可以给一个 Web 浏览器发送多个 cookie，一个 Web 浏览器也可以存储多个 Web 站点提供的 cookie。

（3）浏览器一般只允许存放 300 个 cookie，每个站点最多存放 20 个 cookie，每个 cookie 的大小限制为 4 KB。

（4）如果创建了一个 cookie，并将它发送到浏览器，在默认情况下它是一个会话级别的 cookie（即存储在浏览器的内存中），用户退出浏览器之后即被删除。若希望浏览器将该 cookie 存储在磁盘上，则需要使用 maxAge，并给出一个以秒为单位的时间。将最大时效设为 0 则是命令浏览器删除该 cookie。

5.2.5 删除 cookie

注意：删除 cookie 时，path 必须一致，否则不会删除。

```
// 将 cookie 的有效期设置为 0，命令浏览器删除该 cookie
 cookie.setMaxAge(0);
```

5.2.6 cookie 中存取中文

要想在 cookie 中存储中文，那么必须使用 URLEncoder 类里面的 encode(String s, String enc)方法进行中文转码，例如：

```
 Cookie cookie = new Cookie("userName", URLEncoder.encode(" 融创软通 IT 学院 ",
"UTF-8"));
 response.addCookie(cookie);
```

在获取 cookie 中的中文数据时，再使用 URLDecoder 类里面的 decode(String s, String enc)进行解码，例如：

```
URLDecoder.decode(cookies[i].getValue(), "UTF-8")
```

5.3 使用 JavaMail 发送和接收邮件

现在很多网站都有用户注册功能，通常我们注册成功之后就会收到一封来自注册网站的邮件。邮件里面的内容可能包含了我们注册的用户名和密码以及一个激活账户的超链接等信息。今天我们也来实现一个这样的功能，用户注册成功之后，就将用户的注册信息以 E-mail 的形式发送到用户的注册邮箱当中，实现发送邮件功能需要借助于 JavaMail，需要下载 activation.jar 和 mail.jar 或配置 pom.xml。

配置文件中的部分代码如下。

```
<dependency>
            <groupId>javax.mail</groupId>
            <artifactId>mail</artifactId>
            <version>1.4</version>
</dependency>
```

下载成功的依赖包如图 5.3 所示。

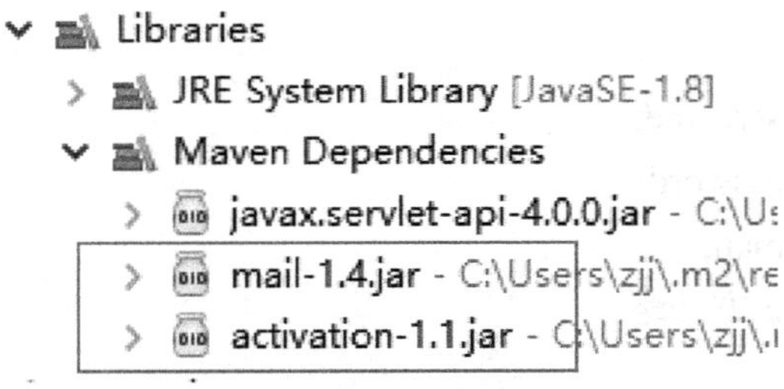

```
Libraries
  > JRE System Library [JavaSE-1.8]
  Maven Dependencies
    > javax.servlet-api-4.0.0.jar - C:\U
    > mail-1.4.jar - C:\Users\zjj\.m2\re
    > activation-1.1.jar - C:\Users\zjj\.
```

图 5.3　项目所用包

5.3.1 用户注册的 JSP 页面

register.jsp 代码如下。

```
<body>
    <form action="registerServlet" method="post">
            用户名:<input type="text" name="username"><br />
            密      码: <input type="password" name="pass-
word"><br />
            邮      箱:<input type="text" name="email"><br />
            <input type="submit" value=" 注册 ">
```

```
    </form>
</body>
```

5.3.2 消息提示页面

message.jsp 代码如下。

```
<%@ page language="java" pageEncoding="UTF-8"%>
<!DOCTYPE HTML>
<html>
<head>
<title> 消息提示页面 </title>
</head>
<body>${message}
</body>
</html>
```

5.3.3 编写处理用户注册处理程序

编写封装用户注册信息的 pojo：User.java 代码如下。

```
public class User {
    private String username;
    private String password;
    private String email;
    // setter 和 getter 方法略
}
```

5.3.4 编写邮件发送功能

提示：发送邮件是一件非常耗时的事情，因此一定要用一个线程类来发送邮件。

```
package com.isoft.servlet;
// 导入包略
public class SendMail extends Thread {
    // 用于给用户发送邮件的邮箱
    private String from = "rcrt@sohu.com";
    // 邮箱的用户名
    private String username = "rcrt";
    // 邮箱的密码
```

```
    private String password = " 邮箱密码 ";
    // 发送邮件的服务器地址
    private String host = "smtp.sohu.com";
    private User user;
    public SendMail(User user) {
        this.user = user;
    }
    /*
     * 重写 run 方法的实现,在 run 方法中发送邮件给指定的用户
     */
    @Override
    public void run() {
        try {
            Properties prop = new Properties();
            prop.setProperty("mail.host", host);
            prop.setProperty("mail.transport.protocol", "smtp");
            prop.setProperty("mail.smtp.auth", "true");
            Session session = Session.getInstance(prop);
            session.setDebug(true);
            Transport ts = session.getTransport();
            ts.connect(host, username, password);
            Message message = createEmail(session, user);
            ts.sendMessage(message, message.getAllRecipients());
            ts.close();
        } catch (Exception e) {
            throw new RuntimeException(e);
        }
    }
    /**

     * @Description: 创建要发送的邮件
     */
    public Message createEmail(Session session, User user) throws Exception {
        MimeMessage message = new MimeMessage(session);
        message.setFrom(new InternetAddress(from));
        message.setRecipient(Message.RecipientType.TO, new InternetAddress(user.
getEmail()));
```

```
            message.setSubject(" 用户注册邮件 ");
            String info = " 恭喜您注册成功,您的用户名: " + user.getUsername() + ", 您的密码:" + user.getPassword() + ",请妥善保管,如有问题请联系网站客服 !!";
            message.setContent(info, "text/html;charset=UTF-");
            message.saveChanges();
            return message;
        }
    }
```

5.3.5 编写处理用户注册的 Servlet

```
    @WebServlet("/registerServlet")
    public class RegisterServlet extends HttpServlet {
        public void doGet(HttpServletRequest request, HttpServletResponse response) throws ServletException, IOException {
            try {
                String username = request.getParameter("username");
                String password = request.getParameter("password");
                String email = request.getParameter("email");
                User user = new User();
                user.setEmail(email);
                user.setPassword(password);
                user.setUsername(username);
                System.out.println(" 把用户信息注册到数据库中 ");
                // 用户注册成功之后就向用户注册时的邮箱发送一封 E-mail
                // 发送邮件是一件非常耗时的事情,因此这里开辟了另一个线程来专门发送邮件
                SendMail send = new SendMail(user);
                // 启动线程,线程启动之后就会执行 run 方法来发送邮件
                send.start();
                request.setAttribute("message", " 恭喜您,注册成功,我们已经发了一封带了注册信息的电子邮件,请查收,如果没有收到,可能是网络原因,过一会儿就收到了!! ");
                request.getRequestDispatcher("/message.jsp").forward(request, response);
            } catch (Exception e) {
                e.printStackTrace();
```

```
                request.setAttribute("message", " 注册失败!! ");
                request.getRequestDispatcher("/message.jsp").forward(request,      re-
sponse);
            }
        }
        public void doPost(HttpServletRequest request, HttpServletResponse response) throws
ServletException, IOException {
            doGet(request, response);
        }
    }
```

运行结果如图 5.4 所示。

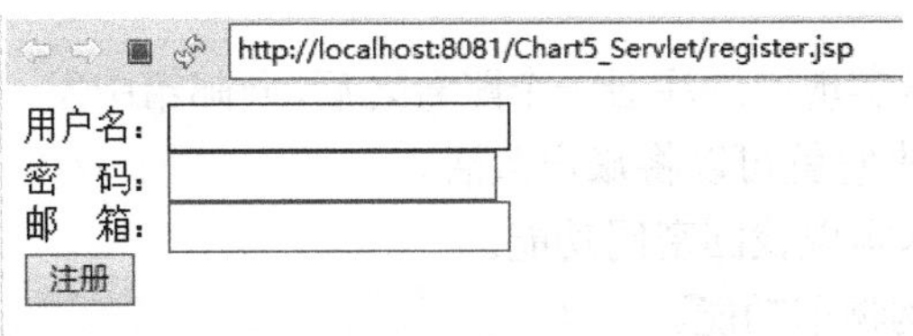

图 5.4　运行结果

现在很多网站都有这样的功能，用户注册完成之后，网站根据我们注册时填写的邮箱给我们发一封 E-mail，然后单击 E-mail 中的超链接去激活我们的用户。

在总结使用 JavaMail 发送邮件时发现，将邮件发送到 sina 或者 sohu 的邮箱时，不一定能够马上收取得到邮件，总是有延迟，有时甚至会延迟很长的时间，甚至会被当成垃圾邮件来处理掉，或者干脆就拒绝接收，有时候为了看到邮件发送成功的效果，要等很长时间。

小结

本章主要讲解了 3 个方面的内容。

(1)通过一个 Servlet 接收多个用户请求的两种方法(使用传递参数的方法和使用反射机制)。

(2)详细讲解了 cookie 在 Servlet 中的应用，cookie 是一个非常有用的一个简单应用，主要负责在客户端存储数据。

(3)讲解了 JavaMail 的简单应用，如果想深入理解 JavaMail 的原理和使用，还需要进行深入学习。

经典面试题

1. 如何使一个 Servlet 处理多个请求？

2.Servlet 如何使用 cookie？

3. 如何使用 cookie 实现记住密码功能？

4. 如何解决 Servlet 读取 cookie 中文乱码的问题？

5.Servlet 如何删除 cookie？

跟我上机

1.Servlet 方式通过 cookie 记住登录时的用户名和密码。

2. 使用 cookie 实现访问购物网站的某个商品以后，下次继续来访问这个网站，会有一个上次浏览物品的显示。

3. 使用 JavaMail 技术实现网站注册一个账号之后，向邮箱中发一封账户激活的邮件，邮件中有一个激活的链接，单击它就可以将账户激活。

4. 使用 JavaMail 技术实现找回密码功能。

5. 使用 Servlet 实现购物车功能。

第 6 章　JSP 技术

本章要点（学会后请在方框里打钩）：

- □ 了解 JSP 是什么
- □ 了解 JSP 的工作原理
- □ 掌握 JSP 的语法基础
- □ 了解 JSP 的元素构成
- □ 熟练使用 Eclipse 和 Intellij IDEA 开发 JSP

6.1 JSP 技术概述

JSP 的全称是 Java Server Pages（Java 服务器页面），和 Servlet 技术一样，都是 SUN 公司定义的一种用于开发动态 Web 资源的技术，当前的最高版本为 2.3。

JSP 这门技术的最大特点在于写 JSP 就像在写 HTML，但与 HTML 只能为用户提供静态数据相比，JSP 技术允许在页面中嵌套 Java 代码，为用户提供动态数据。请注意以下区别。

（1）JSP（Java 服务器页面）以扩展名 .jsp 保存。

（2）有效控制动态内容的生成。

（3）使用 Java 编程语言和 Java 类库。

（4）HTML 用于表示页面，而 Java 代码用于访问动态内容。

专家讲解

Servlet 与 JSP 的区别。

（1）JSP 经编译后就变成了 Servlet（JSP 的本质就是 Servlet，JVM 只能识别 Java 的类，不能识别 JSP 的代码 ,Web 容器将 JSP 的代码编译成 JVM 能够识别的 Java 类）。

（2）JSP 更擅长页面显示，Servlet 更擅长逻辑控制。

（3）Servlet 中没有内置对象，内置对象都是必须通过 HttpServletRequest 对象实现的，HttpServletResponse 对象以及 HttpServlet 对象得到 JSP 是 Servlet 的一种简化，使用 JSP 只需要完成程序员需要输出到客户端的内容，JSP 中的 Java 脚本镶嵌到一个类中是 JSP 容器完成的。而 Servlet 则是一个完整的 Java 类，这个类的 Service 方法用于生成对客户端的响应。

（4）对于静态 HTML 标签，Servlet 都必须使用页面输出流逐行输出。

6.1.1 JSP 最佳实践

虽然不管是 JSP 还是 Servlet 都可以用于开发动态 Web 资源，但由于这两门技术各自的特点，在长期的软件实践中，人们逐渐把 Servlet 作为 Web 应用中的控制器组件来使用，而把 JSP 技术作为数据显示模板来使用。其原因为，程序的数据通常要美化后再输出。让 JSP 既用 Java 代码产生动态数据，又做美化会导致页面难以维护。让 Servlet 既产生数据，又在里面嵌套 HTML 代码美化数据，同样也会导致程序可读性差，难以维护。因此最好的办法就是根据这两门技术的特点，让它们各司其职，Servlet 只负责响应请求产生数据，并把数据通过转发技术带给 JSP，数据的显示则由 JSP 来做。

6.1.2 JSP 的执行过程

JSP 的执行过程如图 6.1 所示。

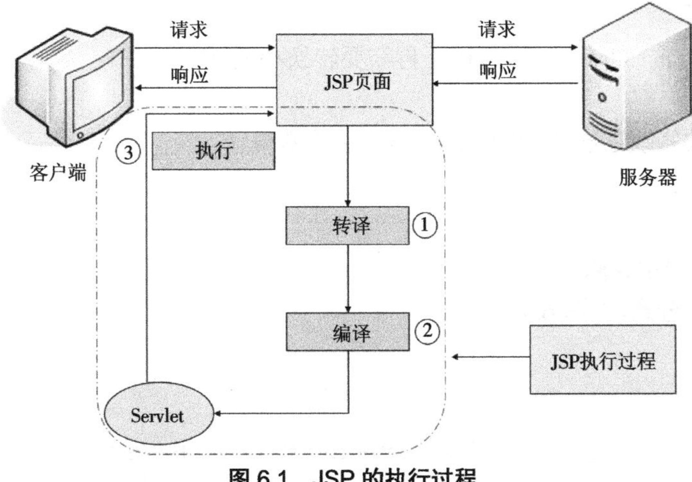

图 6.1　JSP 的执行过程

6.1.3　Tomcat 服务器的执行流程

Tomcat 服务器的执行流程如图 6.2 所示。

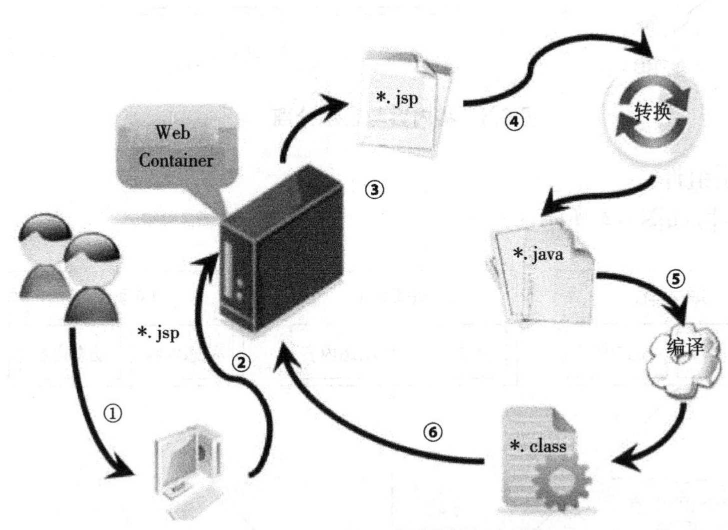

图 6.2　Tomcat 服务器的执行流程

1. 第一次执行

（1）客户端通过计算机连接服务器，因为请求是动态的，所以所有的请求交给 Web 容器（Tomcat）来处理。

（2）在容器中找到需要执行的 *.jsp 文件。

（3）之后将 *.jsp 文件转译为 *.java 文件。

（4）产生的 *.java 文件（就是一个 Servlet）经过编译器编译后，形成 *.class 文件。

（5）最终服务器要执行形成的 *.class 文件。

2. 第二次执行

因为已经存在了 *.class 文件,所以不再需要转换和编译。

3. 修改后执行

源文件已经被修改过了,所以需要重新转换,重新编译。

6.1.4 JSP 的优点

6.1.4.1 将内容与表示分离

将内容与表示分离,如图 6.3 所示。

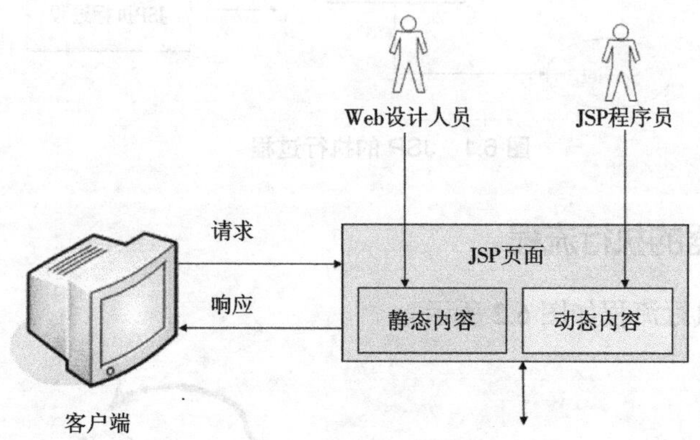

图 6.3 将内容与表示分离

6.1.4.2 强调可重用组件

强调可重用组件,如图 6.4 所示。

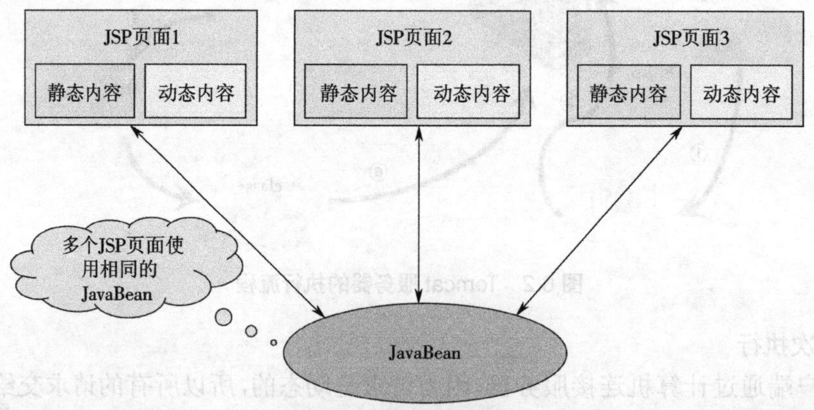

图 6.4 强调可重用组件

6.1.4.3 简化页面开发

简化页面开发,如图 6.5 所示。

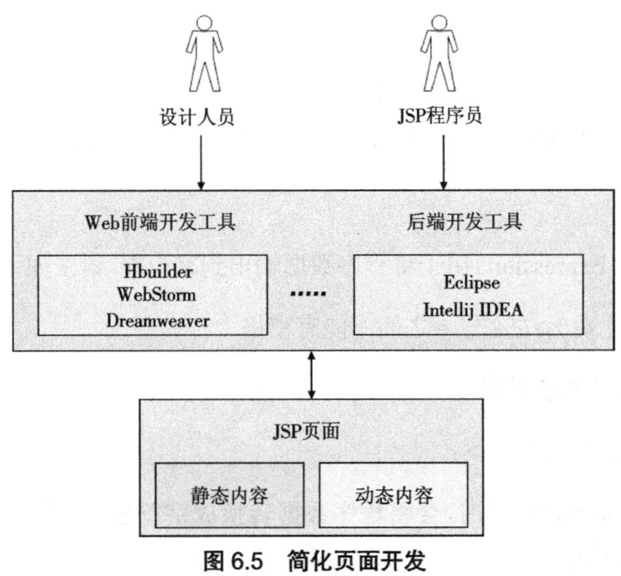

图 6.5　简化页面开发

6.2　JSP 基础语法

任何语言都有自己的语法，JSP 虽然是在 Java 上的一种应用，但是依然有其自己扩充的语法，而且在 JSP 页面中，所有的 Java 语句也都可以使用。图 6.6 是 JSP 页面中能够包含的内容。

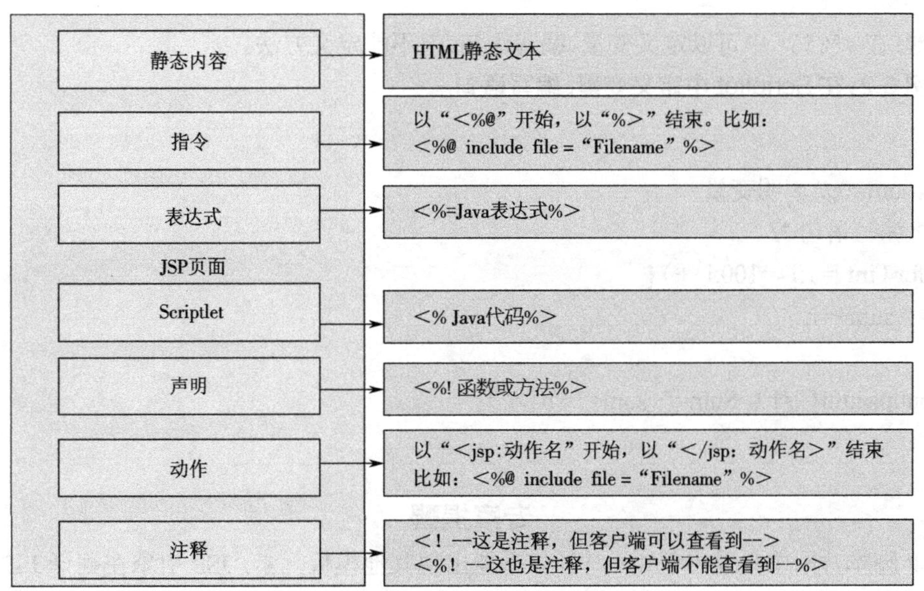

图 6.6　JSP 页面中能够包含的内容

6.2.1 JSP 模版元素

JSP 页面中的 HTML 内容称为 JSP 模版元素。

JSP 模版元素定义了网页的基本骨架，即定义了页面的结构和外观。

6.2.2 JSP 表达式

JSP 脚本表达式（Expression）用于将程序数据输出到客户端，语法如下：

```
<%= 变量或表达式 %>//% 与 = 之间不能有空格
```

实例 6.1：输出当前系统时间

```
<%= new java.util.Date() %>
```

JSP 引擎在翻译脚本表达式时，会将程序数据转换成字符串，然后在相应位置用 out.print(...) 将数据传输给客户端。

提醒：JSP 脚本表达式中的变量或表达式后面不能有分号(;)。

6.2.3 JSP 脚本片断

JSP 脚本片断（也叫 Scriptlet）用于在 JSP 页面中编写多行 Java 代码，语法如下：

```
<%
此处可以写多行 Java 语法代码
%>
```

提醒：在 <% %> 中可以定义变量、编写语句，但不能定义方法。

实例 6.2：在 Scriptlet 中定义变量、编写语句

```
<%
int sum=0;// 声明变量
 /* 编写语句 */
  for (int i=1;i<=100;i++) {
      sum+=i;
  }
 out.println("<h1>Sum="+sum+"</h1>");
%>
```

专家提醒

JSP 脚本片断中只能出现 Java 代码，不能出现其他模板元素，JSP 引擎在翻译 JSP 页面时，会将 JSP 脚本片断中的 Java 代码原封不动地放到 Servlet 的 _jspService 方法中。

JSP 脚本片断中的 Java 代码必须严格遵循 Java 语法，例如，每执行一条语句后面必须用分号(;)结束。

在一个 JSP 页面中可以有多个脚本片断，在两个或多个脚本片断之间可以嵌入文本、HTML 标记和其他 JSP 元素。

实例 6.3：JSP 页面嵌入 Java 代码

```
<%
  int x = 10;
  out.println(x);
%>
<p> 这是 JSP 页面文本 </p>
<%
  int y = 20;
  out.println(y);
%>
```

多个脚本片断中的代码可以相互访问，犹如将所有的代码放在一对 <%%> 之中。

单个脚本片断中的 Java 语句可以是不完整的，但是多个脚本片断组合后必须是完整的 Java 语句，例如：

```
<%
  for (int i = 1; i < 5; i++) {
%>
<H1> 融创软通 </H1>
<%
  }
%>
```

6.2.4 JSP 声明

JSP 页面中编写的所有代码，默认会转译到 Servlet 的 Service 方法中，而 JSP 声明中的 Java 代码将被翻译到 _jspService 方法的外面，语法如下：

```
<%!
java 代码
%>
```

所以，JSP 声明可用于定义 JSP 页面转换成的 Servlet 程序的静态代码块、成员变量和方法。

多个静态代码块、变量和函数可以定义在一个 JSP 声明中，也可以分别单独定义在多个 JSP 声明中。

JSP 隐式对象的作用范围仅限于 Servlet 的 _jspService 方法，所以在 JSP 声明中不能使用这些隐式对象。

实例 6.4：在 JSP 页面中声名 Java 静态代码块，定义变量、函数

```
<%!
  static {
    System.out.println(" 正在加载 ...!");
  }
  private int globalVar = 0;
  public void jspInit() {
    System.out.println(" 初始化方法 !");
  }
%>
<%!
  public void jspDestroy() {
    System.out.println(" 销毁 JSP 方法 !");
  }
%>
```

6.2.5 JSP 注释

在 JSP 中，注释有以下三大类。

（1）显式注释：直接使用 HTML 风格的注释：<!- - 注释内容 - ->。

（2）隐式注释：直接使用 Java 的注释：//、/*……*/。

（3）JSP 自己的注释：<%- - 注释内容 - -%>。

这三种注释的区别如下。

```
<!-- 这个注释可以看见 -->
<%
  //Java 中的单行注释
  /*
   Java 中的多行注释
  */
%>
<%--JSP 自己的注释 --%>
```

专家讲解

HTML 的注释在浏览器中查看源文件的时候是可以看得到的，而 Java 注释和 JSP 注释在浏览器中查看源文件时是看不到的，这就是这三种注释的区别。

6.3 综合实例:根据半径求圆的周长和面积

```
<%!
  double radius = 6.0;
  private double getRadius() {
    return radius;
  }
  private double getDiameter() {
    return (radius * 2);
  }
  private double getArea() {
    return (3.1415 * radius);
  }
  private double getCircumference() {
    return (3.1415 * (radius * 2));
  }
%>
<h3> 计算圆的周长和面积 </h3>
<hr>
<b> 圆的半径 :</b> <%=radius%> cm<br/>
<b> 直径 :</b> <%=getDiameter()%> cm<br/>
<b> 圆的面积为 :</b> <%=getArea()%>
cm<sup>2</sup><br/>
<b> 圆的周长为 :</b>
<%=getCircumference()%><br/>
<hr>
```

运行结果如图 6.7 所示。

图 6.7 运行结果

小结

1.JSP 页面使用 HTML 显示静态内容，并使用 Java 代码生成动态内容。

2.JSP 页面的元素分为静态内容、JSP 指令、JSP 表达式、JSP Scriptlet 和注释。

3. 可以使用标准开发工具创建 JSP 页面。

4.JSP 使用可重用的跨平台组件（如 JavaBean）。

5.JSP 允许创建自定义标签，使 JSP 开发更容易。

6.JSP 执行过程的各个阶段为转译、编译和执行。

经典面试题

1. 描述一下 JSP 是什么。

2. 列举一下 JSP 的语法规则。

3. JSP 中有几种注释？分别是哪几种？

4. JSP 的常用指令有哪些？

5. 如何在 Eclipse 项目中快速找到一个 JSP 文件？

6. JSP 和 Servlet 的区别是什么？

7. 描述一下 JSP 的工作原理。

8. 如果 JSP 页面中出现乱码问题应如何解决？

9. 列举一下 JSP 的 page 指令中的属性有哪些。

10. JSP 里面怎么声明一个 Java 类？如何使用？

跟我上机

1. 编写一个 JSP 页面，用于显示 1~10 的整数，并求总和。

2. 编写一个程序，用于在网页上分别显示 1~10 各数字的阶乘。

3. 编写一个 JSP 程序，使用 JSP Scriptlet 显示网页上不同颜色的颜色条，显示“绿色”“黑色”“红色”的颜色条。

4. 编写一个 JSP 程序实现自动时间显示功能，显示当前时间（时，分，秒），并不断自动刷新。（提示：页面自动刷新 <meta http-equiv="refresh" content="1;url=currenttime.jsp">）

5. 打印九九乘法表，将数据显示在 table 中。

第7章　JSP 指令

本章要点(学会后请在方框里打钩):

- □ 了解什么是 JSP 指令,有哪些指令
- □ 掌握 JSP Page 指令
- □ 熟悉 JSP Page 指令中的各种属性的用法
- □ 掌握 JSP Include 指令的使用
- □ 了解 JSP taglib 指令
- □ 熟悉 Page 指令的 errorpage 属性的使用

7.1 JSP 指令简介

JSP 指令(Directive)是为 JSP 引擎而设计的,它们不直接产生任何可见输出,只是告诉引擎如何处理 JSP 页面中的其余部分。

在 JSP 2.0 规范中共定义了三个指令,如图 7.1 所示。

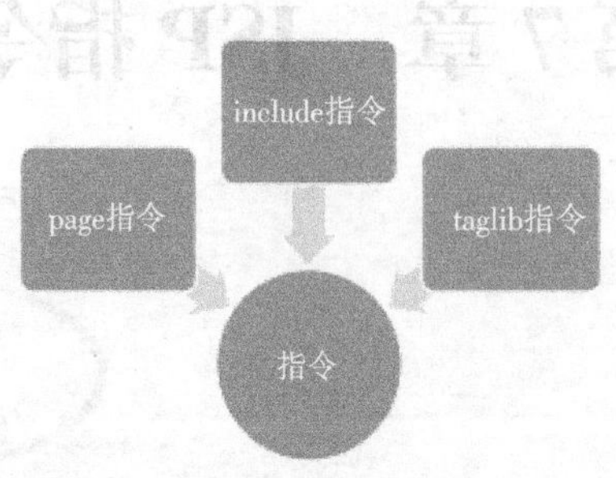

图 7.1 JSP 2.0 规范中定义的指令

JSP 指令的基本语法格式:

```
<%@ 指令 属性名 =" 值 " …%>
```

例如:

```
<%@ page contentType="text/html;charset=UTF-8"%>
```

如果一个指令有多个属性,这些属性可以写在一个指令中,也可以分开写。例如:

```
<%@ page contentType="text/html;charset=UTF-8"%>
<%@ page import="java.util.Date"%>
```

也可以写作:

```
<%@ page contentType="text/html;charset=UTF-8" import="java.util.Date"%>
```

提示:不写指令时,在 JSP 页面中 JSP 引擎会自动导入图 7.2 所示的包。

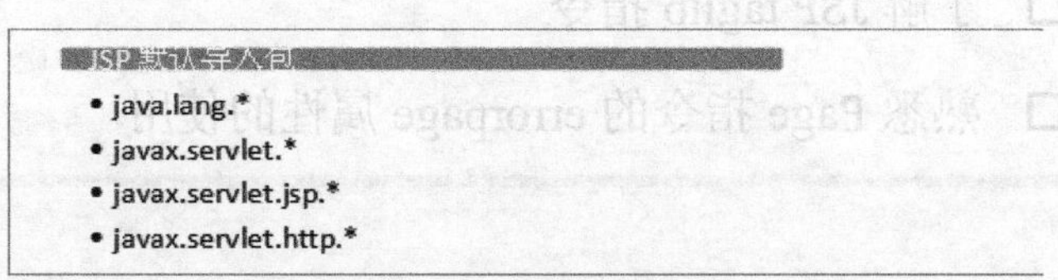

图 7.2 JSP 默认导入包

7.2 page 指令

page 指令用于定义 JSP 页面的各种属性，无论 page 指令出现在 JSP 页面中的什么地方，它作用的都是整个 JSP 页面，为了保持程序的可读性和遵循良好的编程习惯，page 指令最好放在整个 JSP 页面的起始位置（第一行），如图 7.3 所示。

图 7.3 page 指令在 JSP 页面的位置

JSP 2.0 规范中定义的 page 指令的完整语法如下。

```
<%@ page
    [ language="java" ]
    [ extends="package.class" ]
    [ import="{package.class | package.*}, ..." ]
    [ session="true | false" ]
    [ buffer="none | 8kb | sizekb" ]
    [ autoFlush="true | false" ]
    [ isThreadSafe="true | false" ]
    [ info="text" ]
    [ errorPage="relative_url" ]
    [ isErrorPage="true | false" ]
    [ contentType="mimeType [ ;charset=characterSet ]" | "text/html ; charset=ISO-8859-1" ]
    [ pageEncoding="characterSet | ISO-8859-1" ]
    [ isELIgnored="true | false" ]
%>
```

7.2.1 page 指令属性

page 指令可以定义下面这些大小写敏感的属性（大致按照使用的频率列出）：import、contentType、pageEncoding、session、isELIgnored（只限 JSP 2.0）、buffer、autoFlush、info、errorPage、isErrorPage、isThreadSafe、language 和 extends。

7.2.1.1 import 属性

使用 page 指令的 import 属性指定 JSP 页面在转换为 Servlet 类后应导入的依赖包。在 JSP 中，包是绝对必需的。原因是，如果没有使用包，系统即认为所引用的类与当前类在同一个包中。

使用 import 属性时，可以采用下面两种形式：

```
<%@ page import="package.class" %>
<%@ page import="package.class1, ..., package.classN" %>
```

例如，下面的指令表示 java.util 包和 com.isoft 包中的所有类在使用时无须给出明确的包标识符：

```
<%@ page import="java.uti1.*, com.isoft.*" %>
```

import 是 page 的属性中唯一允许在同一文档中多次出现的属性。尽管 page 指令可以出现在文档中的任何地方，但一般不是将 import 语句放在文档顶部附近，就是放在相应的包首次使用之前。

7.2.1.2　contentType 和 pageEncoding 属性

contentType 属性设置 Content-Type 响应报头，标明即将发送到客户程序的文档的 MIME 类型。使用 contentType 属性时，可以采用下面两种形式：

```
<%@ page contentType="MIME-TYPE" %>
<%@ page contentType="MIME-Type; charset=Character-Set" %>
```

例如，指令：

```
<%@ page contentType="application/vnd.ms-excel" %>
```

和下面的 scriptlet 所起到的作用基本相同。

```
<% responce.setContentType("application/vnd.ms-excel"); %>
```

response.setContentType 方法使用明确的 Java 代码（这是一些开发人员力图避免使用的方式），而 page 指令只用到 JSP 语法。

```
<%@ page contentType="someMimeType; charset=someCharacterSet" %>
```

但是，如果只想更改字符集，使用 pageEncoding 属性更为简单。例如，中文 JSP 页面可以使用下面的语句：

```
<%@ page pageEncoding="utf-8" %>
```

7.2.1.3　session 属性

session 属性控制页面是否参与 HTTP 会话。使用这个属性时，可以采用下面两种形式：

```
<%@ page session="true" %> <%--Default--%>
<%@ page session="false" %>
```

true 值（默认）表示：如果存在已有会话，则预定义变量 session（类型为 HttpSession）应该绑定到现有的会话；否则，创建新的会话并将其绑定到 session。false 值表示不自动创建会话，在 JSP 页面转换成 Servlet 时，对变量 session 的访问会产生错误。

对于高流量的网站，使用 session="false" 可以节省大量的服务器内存。但要注意，session="false" 并不禁用会话跟踪，它只是阻止 JSP 页面为那些尚不拥有会话的用户创建新的会话。由于会话是针对用户，不是针对页面，所以关闭某个页面的会话跟踪没有任何益处，除非在同一客户会话中被访问到的相关页面全部被关闭。

7.2.1.4 isELIgnored 属性

这是 JSP 2.0 引入的属性，使用这个属性时，可以采用下面两种形式：

```
<%@ page isELIgnored="true" %>// 忽略 JSP 2.0 表达式语言（EL）
<%@ page isELIgnored="false" %>//EL 表达式可以进行正常的求值
```

请在 JSP 页面中尝试 :${3+5}, 就能看到效果。

7.2.1.5 buffer 和 autoFlush 属性

buffer 属性指定 out 变量（类型为 JspWriter ）使用的缓冲区的大小。使用这个属性时，可以采用下面两种形式：

```
<%@ page buffer="sizekb" %>
<%@ page buffer="none" %>
```

服务器实际使用的缓冲区可能比指定的更大，但不会小于指定的大小。例如，<%@ page buffer="32kb" %> 表示应该对文档的内容进行缓存，除非页面缓存内容大小为 32 KB、页面完成或明确地对输出执行清空（例如使用 response. flushBuffer)，否则不将文档发送给客户。

默认的缓冲区大小与服务器相关，但至少为 8 KB。如果要将缓冲功能关闭，应该十分小心：这样做要求设置报头或状态代码的 JSP 元素都要出现在文件的顶部，位于任何 HTML 内容之前。另一方面，有时输出内容的每一行都需要较长的生成时间，此时禁用缓冲或使用小缓冲区会更有效率，这样用户能够在每一行生成之后立即看到它们，而不是等待更长的时间看到成组的行。

autoFlush 属性如果是 true，当缓冲区充满之后，应该自动清空输出缓冲区（默认），如果是 false，则缓冲区溢出后抛出一个异常。使用这个属性时，可以采用下面两种形式：

```
<%@ page autoFlush="true" %> <%--Default--%>
< %@ page autoFlush="false" %>
```

在 buffer="none" 时，false 值是不合法的。如果客户程序是常规的 Web 浏览器，那么 autoFlush="false" 的使用极为罕见。但是，如果客户程序是定制应用程序，可能希望确保应用程序要么接收到完整的消息，要么根本没有消息。false 值还可以用来捕获产生过多数据的数据库查询，但是一般说来，将这些逻辑放在数据访问代码中（而非表示代码）要更好一些。

7.2.1.6 info 属性

info 属性可以定义一个在 Servlet 中通过 getServletInfo 方法获取的字符串，使用 info 属性时，采用下面的形式：

```
<%@ page info="Some Message" %>
```

7.2.1.7 errorPage 和 isErrorPage 属性

errorPage 属性用来指定一个 JSP 页面，由该页面来处理当前页面中抛出但未被捕获的任何异常（即类型为 Throwable 的对象）。它的应用方式如下：

```
<%@ page errorPaqe="Relative URL" %>
```

指定的错误页面可以通过 exception 变量抛出的异常。

isErrorPage 属性表示当前页是否可以作为其他 JSP 页面的错误页面。使用 isErrorPage 属性时，可以采用下面两种形式：

```
<%@ page isErrorPage="true" %>//exception 对象存在
< %@ page isErrorPage="false" %> <%--Default--%>//exception 对象不存在
```

7.2.1.8 isThreadSafe 属性

isThreadSafe 属性控制由 JSP 页面生成的 Servlet 是允许并行访问（默认），还是同一时间不允许多个请求访问单个 Servlet 实例（isThreadSafe="false"）。使用 isThreadSafe 属性时，可以采用下面两种形式：

```
<%@ page isThreadSafe="true" %> <%--Default--%>
< %@page isThreadSafe="false" %>
```

遗憾的是，阻止并发访问的标准机制是实现 SingleThreadModel 的接口。尽管在早期推荐使用 SingleThreadModel 和 isThreadSafe="false"，但最近的经验表明 SingleThreadModel 的设计很差，使得它基本毫无用处。因而，应该避免使用 isThreadSafe，采用显式的同步措施取而代之。

7.2.1.9 extends 属性

extends 属性指定 JSP 页面生成的 Servlet 的超类（superclass）。它采用下面的形式：

```
<%@ page extends="package.class" %>
```

这个属性一般为开发人员或提供商保留，由他们对页面的运作方式做出根本性的改变（如添加个性化特性）。一般人应该避免使用这个属性，除非引用由服务器提供商专为这种目的提供的类。

7.2.1.10 language 属性

从某种角度讲，language 属性的作用是指定页面使用的脚本语言，如下所示：

```
<%@ page language="java" %>
```

就现在来说，由于 Java 既是默认选择，也是唯一合法的选择，所以没必要再去关心这个属性。

7.2.2 page 指令中的 errorPage 属性的使用

（1）errorPage 属性的设置值必须使用相对路径，如果以“/”开头，表示相对于当前 Web 应用程序的根目录（注意不是站点根目录），否则表示相对于当前页面。

（2）可以在 web.xml 文件中使用 <error-page> 元素为整个 Web 应用程序设置错误处理页面，具体内容如下。

① <error-page> 元素有 3 个子元素，<error-code>、<exception-type>、<location>。

② <error-code> 子元素指定错误的状态码，例如：<error-code>404</error-code>。

③ <exception-type> 子元素指定异常类的完全限定名，例如：<exception-type>java.lang.ArithmeticException</exception-type>。

④ <location> 子元素指定以"/"开头的错误处理页面的路径，例如：<location>/ErrorPage/404Error.jsp</location>。

专家提醒

如果设置了某个 JSP 页面的 errorPage 属性，那么在 web.xml 文件中设置的错误处理将不对该页面起作用。

7.2.2.1 使用 errorPage 属性指明出错后跳转的错误页面

test.jsp 页面的代码如下。

```
<%@ page language="java" import="java.util.*" errorPage="/ErrorPage/error.jsp" pageEncoding="UTF-8" %>
<html>
<head>
  <title> 测试 page 指令的 errorPage 属性 </title>
</head>
<body>
<%
  // 这行代码肯定会出错，因为除数是 0，一运行就会抛出异常
  int x = 1 / 0;
%>
</body>
</html>
```

在 test.jsp 中，page 指令的 errorPage 属性指明了出错后跳转到 "/ErrorPage/error.jsp"，error.jsp 页面的代码如下：

```
<%@ page language="java" import="java.util.*" pageEncoding="UTF-8" %>
<html>
<head>
  <title> 错误信息友好提示页面 </title>
</head>
<body>
对不起，出错了，请联系管理员解决！
```

```
</body>
</html>
```

运行结果如图 7.4 所示。

图 7.4　运行结果

7.2.2.2　web.xml 中配置 <error-page>

实例 7.1：使用 <error-page> 标签配置针对 404 错误的处理页面

web.xml 的代码如下。

```
<?xml version="1.0" encoding="UTF-8"?>
<web-app xmlns="http://xmlns.jcp.org/xml/ns/javaee"
    xmlns:xsi="http://www.w3.org/2001/XMLSchema-instance"
     xsi:schemaLocation="http://xmlns.jcp.org/xml/ns/javaee http://xmlns.jcp.org/xml/ns/ja-
vaee/web-app_3_1.xsd"
     version="3.1">
  <error-page>
    <location>/404Error.jsp</location>
    <error-code>404</error-code>
  </error-page>
</web-app>
```

404Error.jsp 的代码如下。

```
<%@ page language="java" import="java.util.*" pageEncoding="UTF-8" %>
<html>
<head>
  <title>404 错误友好提示页面 </title>
  <!-- 3 秒钟后自动跳转回首页 -->
    <meta  http-equiv="refresh"  content="3;url=${pageContext.request.contextPath}/index.
jsp">
</head>
<body>
<img alt=" 对不起，你要访问的页面没有找到，请联系管理员处理 !"
   src="${pageContext.request.contextPath}/img/404Error.png"/><br/>
```

```
    3 秒钟后自动跳转回首页，如果没有跳转，请点击 <a href="${pageContext.request.contextPath}/index.jsp"> 这里 </a>
    </body>
    </html>
```

当访问一个不存在的 Web 资源时，就会跳转到在 web.xml 中配置的 404 错误处理页面 404Error.jsp，如图 7.5 所示。

localhost:8088/index1.jsp　没有这个资源时

对不起，你要访问的页面没有找到，请联系管理员处理!

3秒钟后自动跳转回首页，如果没有跳转，请点击这里

图 7.5　请求所访问页面不存在，跳转到提前写好的错误页面

7.2.2.3　使用 page 指令的 isErrorPage 属性显式声明页面为错误处理页面

如果某个 JSP 页面是作为系统的错误处理页面，那么建议将 page 指令的 isErrorPage 属性（默认为 false）设置为“true”来显式声明这个 JSP 页面是一个错误处理页面。

实例 7.2：将 error.jsp 页面显式声明为错误处理页面

```
<%@ page language="java" import="java.util.*" pageEncoding="UTF-8" isErrorPage="true" %>
<html>
<head>
  <title> 错误信息友好提示页面 </title>
</head>
<body>
对不起，出错了，请联系管理员解决！
<%-- 生成 exception 对象，显示异常消息 --%>
<%=exception.getMessage()%>
</body>
</html>
```

将 error.jsp 页面显式声明为错误处理页面后的好处就是 JSP 引擎在将 JSP 页面翻译成 Servlet 时，在 Servlet 的 _jspService 方法中会声明一个 exception 对象，然后将运行 JSP 出错的异常信息存储到 exception 对象中。

如果没有设置 isErrorPage="true"，那么在 JSP 页面中是无法使用 exception 对象的。

JSP 有九大内置对象，而在一般情况下 exception 对象在 JSP 页面中是获取不到的，只有设置 page 指令的 isErrorPage 属性为“true”来声明 JSP 页面是一个错误处理页面之后才能够在 JSP 页面中使用 exception 对象。

7.3 include 指令

在 JSP 中对于“包含”一个页面有下面两种语句形式。

(1)使用 @include 指令。

(2)使用 <jsp:include> 标准动作。

7.3.1 include 指令

@include 可以包含任意的文件。当然只是把文件的内容包含进来。

include 指令用于引入其他 JSP 页面,如果使用 include 指令引入了其他 JSP 页面,那么 JSP 引擎会把这两个 JSP 翻译成一个 Servlet。所以 include 指令引入通常也称为静态引入。语法如下:

```
<%@ include file="relativeURL"%>
```

其中的 file 属性用于指定被引入文件的路径。路径以“/”开头,表示代表当前 Web 应用。

include 指令细节上应注意问题如下。

(1)被引入的文件必须遵循 JSP 语法。

(2)被引入的文件可以使用任意的扩展名,即使其扩展名是 html, JSP 引擎也会按照处理 JSP 页面的方式处理它里面的内容,为了见名知意, JSP 规范建议使用 .jspf(JSP fragments(片段))作为静态引入文件的扩展名。

(3)由于使用 include 指令将会涉及两个 JSP 页面,并会把两个 JSP 翻译成一个 Servlet,所以这两个 JSP 页面的指令不能冲突(除了 pageEncoding 和导包除外)。

综合实例:新建 head.jspf 页面和 foot.jspf 页面,分别作为 JSP 页面的头部和尾部,存储于 WebRoot 下的 jspfragments 文件夹中

head.jspf 代码如下。

```
<%@ page language="java" import="java.util.*" pageEncoding="UTF-8"%>
<h1 style="color:red;"> 网页头部 </h1>
```

foot.jspf 代码如下。

```
<%@ page language="java" import="java.util.*" pageEncoding="UTF-8"%>
<h1 style="color:blue;"> 网页尾部 </h1>
```

在 WebRoot 文件夹下创建一个 includeTagTest.jsp 页面,在 includeTagTest.jsp 页面中使用 @include 指令引入 head.jspf 页面和 foot.jspf 页面,代码如下:

```
<%@ page language="java" import="java.util.*" pageEncoding="UTF-8"%>
<html>
<head>
   <title>JSP 的 Include 指令测试 </title>
</head>
<body>
  <%-- 使用 include 标签引入引入其他 JSP 页面 --%>
   <%@include file="/jspfragments/head.jspf" %>
   <h1> 网页主体内容 </h1>
   <%@include file="/jspfragments/foot.jspf" %>
  </body>
 </html>
```

专家讲解

使用 @include 可以包含任意的内容，文件的后缀是什么都无所谓。这种把别的文件内容包含到自身页面的 @include 语句就叫作静态包含，作用只是把别的页面内容包含进来，属于静态包含。

7.3.2 <jsp:include> 标准动作

jsp:include 标准动作为动态包含，如果被包含的页面是 JSP，则先处理之后再将结果包含，而如果包含的是非 *.jsp 文件，则只是把文件内容静态包含进来，功能与 @include 类似。后面再具体介绍。

7.4 taglib 指令

提示：本章只是简单说明一下使用方式，具体细节使用方法将在第 11 章中介绍。

语法格式：

```
<%@ taglib uri=" 引用标签的路径 "    prefix=" 引用标签的前缀 "%>
```

7.4.1 自定义标签使用

可以自己起一个有个性的名字，但这样做的后果就是编译器找不到用标签的 tld 文件，从而找不到这个标签，最终导致页面无法正常工作。

JSP 页面使用自定义标签库，代码如下。

```
<%@ page language="java" contentType="text/html; charset=UTF-8"
    pageEncoding="UTF-8"%>
<%@ taglib prefix="c" uri="myjstl" %>
<!DOCTYPE html PUBLIC "-//W3C//DTD HTML 4.01 Transitional//EN" "http://www.w3.org/TR/html4/loose.dtd">
<html>
<head>
    <meta http-equiv="Content-Type" content="text/html; charset=UTF-8">
    <title>Insert title here</title>
</head>
<body>
    <c:out value="${param.username}"/>
</body>
</html>
```

如果使用了自定义 uri，还需要在该工程的 web.xml 下加入如下信息。

```
<web-app id="WebApp_ID" version="2.4" xmlns="http://java.sun.com/xml/ns/j2ee" xmlns:xsi="http://www.w3.org/2001/XMLSchema-instance" xsi:schemaLocation="http://java.sun.com/xml/ns/j2ee http://java.sun.com/xml/ns/j2ee/web-app_2_4.xsd">
        ...
    <jsp-config>
      <taglib>
          <taglib-uri>myjstl</taglib-uri>
          <taglib-location>/WEB-INF/tld/c.tld</taglib-location>
      </taglib>
    </jsp-config>
        ...
</web-app>
```

提示：c.tld 必须存在或者自己编写。

7.4.2 标准标签的使用

引入标签库很简单，先将标签库 jar 文件放到工程项目中，然后实现下面页面。

实例 7.3：使用迭代标签(foreach)遍历数据

```
<%@ page language="java" contentType="text/html; charset=utf-8"
    pageEncoding="utf-8"%>
```

```
<%@ taglib uri="http://java.sun.com/jsp/jstl/core" prefix="c" %>
<!DOCTYPE html PUBLIC "-//W3C//DTD HTML 4.01 Transitional//EN" "http://www.w3.org/TR/html4/loose.dtd">
<html>
<head>
<meta http-equiv="Content-Type" content="text/html; charset=utf-8">
<title>collection</title>
</head>
<body>
<form action="/test/CollectionServlet" method="post">
<p><input name="name" size="20" value=""></p>
<p><input name="url" size="50" value=""></p>
<p><input name="" type="submit" value="save"></p>
</form>
<hr>
${q }
<c:forEach var="l" items="${list}">
<table>
  <th>
    <td> 编号 </td>
    <td> 名称 </td>
    <td>URL</td>
    <td> 时间 </td>
  </th>
  <tr>
    <td>${l.id}</td>
    <td><a href="${l.url}">${l.name}</a></td>
    <td>${l.createTime}</td>
    <td><a href="${l.url}"> 编辑 </a><a href="${l.url}"> 删除 </a></td>
  </tr>
</table>
<p>${l.url }</p>
</c:forEach>
</body>
</html>
```

提示：本实例在本章仅供参考，具体内容将在第 11 章中详细介绍。

小结

本章主要内容总结如下。

（1）JSP 指令共有三种类型，即 page、include 和 taglib。

（2）JSP 指令包括在 <%@ 和 %> 内。

（3）page 指令用于设置 JSP 页面的属性。

（4）include 指令用于在 JSP 页面嵌入其他文件 。

（5）taglib 指令用于在 JSP 页面中使用标签。

（6）强调使用 errorPage 属性的使用。

经典面试题

1.JSP 指令有哪些？

2.JSP 的 page 指令有哪些属性？

3.JSP 中的指令元素 include 与动作包含 include 有何区别？

4.JSP 页面如何嵌入另一个页面？

5. 如何设置错误页？

跟我上机

1. 以直角三角形的形式显示数字 1~9。

2. 实现一个把整数金额自动转换为带两位小数的金额，使之更加符合人民币的表示习惯。

3. 编写一个 JSP 页面，实现根据一个人的 18 位身份证号显示出其生日的功能，要求把表达式、声明和 Scriptlet 全部用到，并把结果显示在表格中。

身份证	生　日
010030198810092211	1988-10-09

4. 回文字符串的判断。

字符串	是否是回文
aba	True
Abab	False
lovevol	True

5. 张三和李四分别有 800 美元和 1 860 美元，要求把他们的美元换算成人民币，并用表格的形式显示出来，汇率为 8.11。

第 8 章　JavaBean 和标准动作

本章要点(学会后请在方框里打钩):

- □ 了解什么是 JavaBean
- □ 掌握 JSP 中如何使用 JavaBean
- □ 掌握 JSP 中常用标准动作的用法

JavaBean 是一种可重复使用，且跨平台的软件组件。JavaBean 可分为两种：一种是有用户界面（UI）的 JavaBean；还有一种是没有用户界面主要负责处理事务（如数据运算、操纵数据库）的 JavaBean。JSP 通常访问的是后一种 JavaBean。

JSP 标准动作是内置标记，是规范结构，由容器实现，运行时就具有这些功能，每个标准动作能实现一定的功能。

8.1 什么是 JavaBean

JavaBean 是一个遵循特定写法的 Java 类，它通常具有如下特点。

（1）这个 Java 类必须具有一个无参的构造函数。

（2）属性必须私有化。

（3）私有化的属性必须通过 public 类型的方法暴露给其他程序，并且方法的命名也必须遵守一定的命名规范。

实例 8.1：JavaBean

```
/**
 * Person 类就是一个最简单的 JavaBean
 */
public class Person {
    //Person 类封装的私有属性
    private String name;
    private String sex;
    private int age;
    private boolean married;
    /**
     * 无参数构造方法
     */
    public Person() {
    }
//Person 类对外提供的用于访问私有属性的 public 方法
    public String getName() {
        return name;
    }
    public void setName(String name) {
        this.name = name;
    }
    // 其他属性的 getter 和 setter 略
```

```
}
```

JavaBean 在 Java EE 开发中通常用于封装数据，对于遵循以上写法的 JavaBean 组件，其他程序可以通过反射技术实例化 JavaBean 对象，并且通过反射那些遵守命名规范的方法，从而获知 JavaBean 的属性，进而调用其属性保存数据。

8.2 JavaBean 的属性

JavaBean 的属性可以是任意类型，并且一个 JavaBean 可以有多个属性。每个属性通常都需要具有相应的 setter、getter 方法。setter 方法称为属性修改器，getter 方法称为属性访问器。

属性修改器必须以小写的 set 前缀开始，后跟属性名，且属性名的第一个字母大写，例如，name 属性的修改器名称为 setName，password 属性的修改器名称为 setPassword。

属性访问器通常以小写的 get 前缀开始，后跟属性名，且属性名的第一个字母也要改为大写。例如，name 属性的访问器名称为 getName，password 属性的访问器名称为 getPassword。

一个 JavaBean 的某个属性也可以只有 set 方法或 get 方法，这样的属性通常也称为只写、只读属性。

8.3 JSP 和 JavaBean 搭配使用的优点

JSP 和 JavaBean 搭配使用的优点主要有以下几种。

（1）使得 HTML 与 Java 程序分离，这样便于维护代码。如果把所有的程序代码都写到 JSP 网页中，会使代码繁杂，难以维护。

（2）可以降低开发 JSP 网页人员对 Java 编程能力的要求。

（3）JSP 侧重于生成动态网页，事务处理由 JavaBean 来完成，这样可以充分利用 JavaBean 组件的可重用性特点，提高开发网站的效率。

JavaBean 的三个特征如图 8.1 所示。

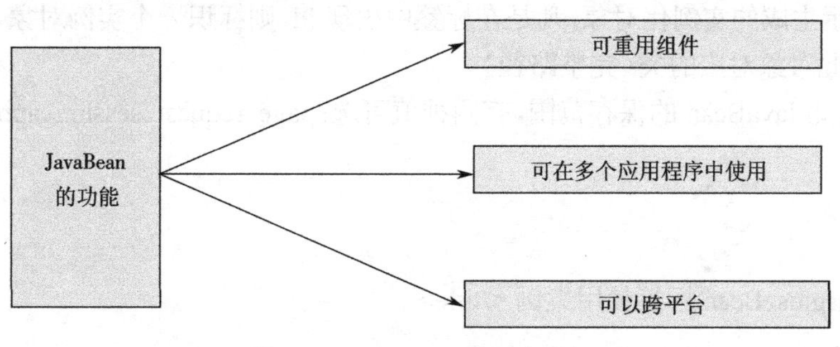

图 8.1 JavaBean 的三个特征

8.4 在 JSP 中使用 JavaBean 的标准动作

JSP 技术提供了三个关于使用 JavaBean 组件的动作元素，即 JSP 标准动作标签，它们分别为如下：

（1）<jsp:useBean> 标签：用于在 JSP 页面中查找或实例化一个 JavaBean 组件。

（2）<jsp:setProperty> 标签：用于在 JSP 页面中设置一个 JavaBean 组件的属性。

（3）<jsp:getProperty> 标签：用于在 JSP 页面中获取一个 JavaBean 组件的属性。

8.4.1 <jsp:useBean> 标签

<jsp:useBean> 标签用于在指定的域范围内查找指定名称的 JavaBean 对象，如果存在则直接返回该 JavaBean 对象的引用，如果不存在则实例化一个新的 JavaBean 对象并将它以指定的名称存储到指定的域范围中。

常用语法如下：

```
<jsp:useBean id="beanName" class="package.class" scope="page|request|session|application"/>
```

“id”属性用于指定 JavaBean 实例对象的引用名称和其存储在域范围中的名称。

“class”属性用于指定 JavaBean 的完整类名（即必须带有包名）。

“scope”属性用于指定 JavaBean 实例对象所存储的域范围，其取值只能是 page、request、session 和 application 四个值中的一个，默认值是 page。

实例 8.2：<jsp:useBean> 标签的使用

```
<%@ page language="java" import="java.util.*" pageEncoding="UTF-8"%>
<jsp:useBean id="person" class="com.iss.bean.Person" scope="page"/>
<%--
    <jsp:useBean>：表示在 JSP 中要使用 JavaBean。
    id: 表示生成的实例化对象，凡是在标签中出现 id，则标识一个实例对象。
    class：此对象对应的类（完整路径）
    scope：此 JavaBean 的保存范围，有四种值可选：page、request、session、application
--%>
<html>
  <head>
    <title>jsp:useBean 标签使用案例 </title>
  </head>
  <body>
```

```
<jsp:useBean id="person" class="com.iss.bean.Person" scope="page"/>
<%
    //person 对象在上面已经使用 jsp:useBean 标签实例化，因此在这里可以直接使用
person 对象
   // 使用 setXxx 方法为对象的属性赋值
   // 为 person 对象的 name 属性赋值
   person.setName（" 融创软通 "）;
   // 为 person 对象的 Sex 属性赋值
   person.setSex（" 男 "）;
   // 为 person 对象的 Age 属性赋值
   person.setAge（24）;
   // 为 person 对象的 married 属性赋值
   person.setMarried（false）;
%>
<%-- 使用 getXxx（）方法获取对象的属性值 --%>
 <h2> 姓名：<%=person.getName（）%></h2>
 <h2> 性别：<%=person.getSex（）%></h2>
 <h2> 年龄：<%=person.getAge（）%></h2>
 <h2> 已婚：<%=person.isMarried（）%></h2>
</body>
</html>
```

运行结果如图 8.2 所示。

localhost:8088/index.jsp

姓名：融创软通

性别：男

年龄：24

已婚：false

图 8.2 运行结果

专家讲解

<jsp:useBean> 标签的执行原理：首先在指定的域范围内查找指定名称的 JavaBean 对象，如果存在则直接返回该 JavaBean 对象的引用，如果不存在则实例化一个新的 JavaBean 对象并将它以指定的名称存储到指定的域范围中。

8.4.2 带标签体的 <jsp:useBean> 标签

带标签体的 <jsp:useBean> 标签语法：

```
<jsp:useBean ...>
Body
</jsp:useBean>
```

Body 部分的内容只在 <jsp:useBean> 标签创建 JavaBean 的实例对象时才执行。这种做法用的不多，了解一下即可。

8.4.3 <jsp:setProperty> 标签

<jsp:setProperty> 标签用于设置和访问 JavaBean 对象的属性。

语法格式一：

```
<jsp:setProperty name="beanName" property="propertyName" value="string 字符串 "/>
```

语法格式二：

```
<jsp:setProperty name="beanName" property="propertyName" value="<%= expression %>"
/>
```

语法格式三：

```
<jsp:setProperty name="beanName" property="propertyName" param="parameterName"/>
```

语法格式四：

```
<jsp:setProperty name="beanName" property= "*" />
```

name 属性用于指定 JavaBean 对象的名称。

property 属性用于指定 JavaBean 实例对象的属性名。

value 属性用于指定 JavaBean 对象的某个属性的值，value 的值可以是字符串，也可以是表达式。如果 value 的值是一个字符串，该值会自动转化为 JavaBean 属性相应的类型，如果 value 的值是一个表达式，那么该表达式的计算结果必须与所要设置的 JavaBean 属性的类型一致。

param 属性用于将 JavaBean 实例对象的某个属性值设置为一个请求参数值，该属性值同样会自动转换成要设置的 JavaBean 属性的类型。

实例 8.3：使用 jsp:setProperty 标签设置 person 对象的属性值

```
<%@ page language="java" import="java.util.*" pageEncoding="UTF-8" %>
<jsp:useBean id="person" class="com.iss.bean.Person" scope="page"/>
<%--
```

```
    使用 jsp:setProperty 标签设置 person 对象的属性值
    jsp:setProperty 在设置对象的属性值时会自动把字符串转换成八种基本数据类型
    但是 jsp:setProperty 对于复合数据类型无法自动转换
    --%>
    <jsp:setProperty property="name" name="person" value=" 融创软通 "/>
    <jsp:setProperty property="sex" name="person" value=" 男 "/>
    <jsp:setProperty property="age" name="person" value="24"/>
    <jsp:setProperty property="married" name="person" value="false"/>
    <%--
    birthday 属性是一个 Date 类型，这个属于复合数据类型，因此无法将字符串自动转换成
Date，用下面这种写法是会报错的
    <jsp:setProperty property="birthday" name="person" value="1988-05-07"/>
    --%>
    <jsp:setProperty property="birthday" name="person" value="<%=new Date()%>"/>
    <!DOCTYPE HTML>
    <html>
    <head>
      <title>jsp:setProperty 标签使用案例 </title>
    </head>
    <body>
    <%-- 使用 getXxx()方法获取对象的属性值 --%>
    <h2> 姓名：<%=person.getName()%></h2>
    <h2> 性别：<%=person.getSex()%></h2>
    <h2> 年龄：<%=person.getAge()%></h2>
    <h2> 已婚：<%=person.isMarried()%></h2>
    <h2> 出生日期：<%=person.getBirthday().toLocaleString()%></h2>
    </body>
    </html>
```

运行结果如图 8.3 所示。

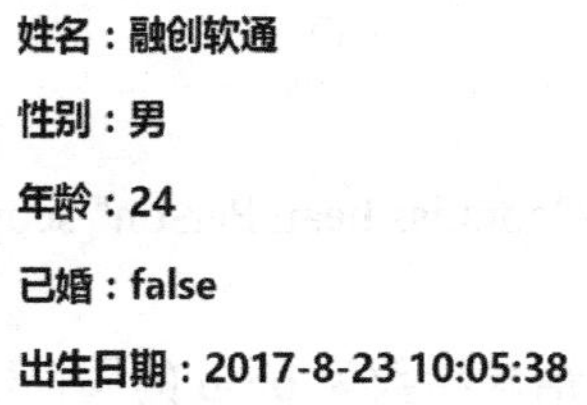

姓名：融创软通

性别：男

年龄：24

已婚：false

出生日期：2017-8-23 10:05:38

图 8.3 运行结果

实例 8.4:使用请求参数为 bean 的属性赋值

```
<%@ page language="java" import="java.util.*" pageEncoding="UTF-8" %>
<!DOCTYPE HTML>
<html>
<head>
  <title>jsp:setProperty 标签使用案例 </title>
</head>
<body>
<jsp:useBean id="person" class="com.iss.bean.Person" scope="page"/>
<%--
  jsp:setProperty 标签可以使用请求参数为 bean 的属性赋值
  param="param_name" 用于接收参数名为 param_name 的参数值,然后将接收到的值赋给 name 属性
--%>
<jsp:setProperty property="name" name="person" param="param_name"/>
<h2> 姓名:<%=person.getName()%></h2>
</body>
</html>
```

运行结果如图 8.4 所示。

← → C ⓘ localhost:8088/index.jsp?param_name=融创软通

姓名：融创软通

图 8.4　运行结果

实例 8.5:用所有的请求参数为 bean 的属性赋值

```
<%@ page language="java" import="java.util.*" pageEncoding="UTF-8" %>
<!DOCTYPE HTML>
<html>
<head>
  <title>jsp:setProperty 标签使用案例 </title>
</head>
<body>
<jsp:useBean id="person" class="com.iss.bean.Person" scope="page"/>
<%--
  jsp:setProperty 标签用所有的请求参数为 bean 的属性赋值
  property="*" 代表 bean 的所有属性
```

```
--%>
<jsp:setProperty property="*" name="person"/>
<h2> 姓名:<%=person.getName()%></h2>
<h2> 性别:<%=person.getSex()%></h2>
<h2> 年龄:<%=person.getAge()%></h2>
</body>
</html>
```

运行结果如图 8.5 所示。

localhost:8088/index.jsp?name=融创软通&sex=女&age=30

姓名：融创软通

性别：女

年龄：30

图 8.5 运行结果

8.4.4 <jsp:getProperty> 标签

<jsp:getProperty> 标签用于读取 JavaBean 对象的属性，也就是调用 JavaBean 对象的 getter 方法，然后将读取的属性值转换成字符串后插入输出的响应正文中。语法如下：

```
<jsp:getProperty name="beanInstanceName" property="PropertyName" />
```

name 属性用于指定 JavaBean 实例对象的名称，其值应与 <jsp:useBean> 标签的 id 属性值相同。

property 属性用于指定 JavaBean 实例对象的属性名。

如果一个 JavaBean 实例对象的某个属性的值为 null，那么，使用 <jsp:getProperty> 标签输出该属性的结果将是一个内容为“null”的字符串。

实例 8.6：使用 jsp:getProperty 获取 bean 对象的属性值

```
<%@ page language="java" import="java.util.*" pageEncoding="UTF-8" %>
<!DOCTYPE HTML>
<html>
<head>
   <title>jsp:getProperty 标签使用案例 </title>
</head>
<body>
<jsp:useBean id="person" class="com.iss.bean.Person" scope="page"/>
<jsp:setProperty property="*" name="person" />
<jsp:setProperty name="person" property="birthday" value="<%=new Date()%>"/>
```

```
<h2> 姓名:<jsp:getProperty property="name" name="person"/></h2>
<h2> 性别:<jsp:getProperty property="sex" name="person"/></h2>
<h2> 年龄:<jsp:getProperty property="age" name="person"/></h2>
<h2> 已婚:<jsp:getProperty property="married" name="person"/></h2>
<h2> 出生日期:<jsp:getProperty property="birthday" name="person"/></h2>
</body>
</html>
```

运行结果如图 8.6 所示。

localhost:8088/index.jsp?name=融创软通&sex=女&age=30

姓名：融创软通

性别：女

年龄：30

已婚：false

出生日期：Wed Aug 23 10:30:07 CST 2017

图 8.6 运行结果

专家讲解

编写 JavaBean 的原则如下。

（1）所有 JavaBean 放在一个包中。

（2）JavaBean 必须声明成 public class，文件名与类名一致。

（3）所有属性必须封装。

（4）设置和取得可以通过 set、get 来实现。

（5）使用 JSP 标签语法取得 JavaBean 必须有一个无参构造方法。

专家讲解

JavaBean 命名规则如下。

（1）包名全部小写。

（2）类名首字母大写。

（3）属性名称，第一个单词首字母小写，之后每个单词首字母大写，如 sayHello。

（4）方法名与属性名相同，如 public void sayHello（）;。

（5）常量名全部大写，如 final String DBDRIVER="com.mysql.jdbc.Driver";。

关于 JavaBean 方面的内容，只需要掌握 JavaBean 的写法以及掌握 <jsp:useBean> 标签、<jsp:setProperty> 标签、<jsp:getProperty> 标签的使用即可。

8.5 JSP 标准动作

JSP 标准动作元素的使用格式为 <jsp: 标记名 >, 它采用严格的 xml 标签语法来表示，这些 JSP 标签动作元素是在用户请求阶段执行的，这些标准动作元素是内置在 JSP 文件中的，所以可以直接使用，不需要进行引用定义，如图 8.7 所示。

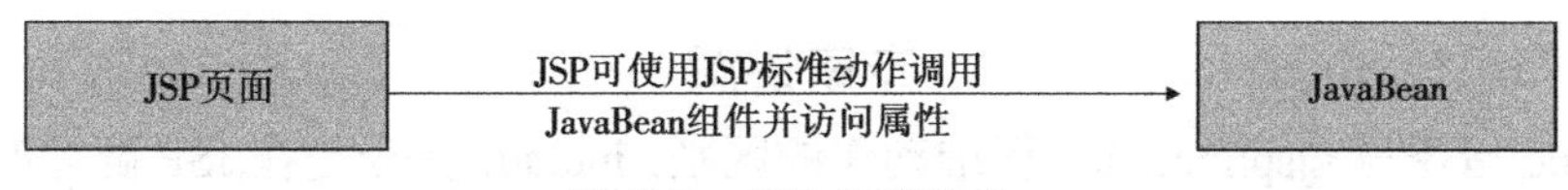

图 8.7 JSP 标准动作

JSP 标准动作元素包括以下几个。

（1）< jsp:useBean >：定义 JSP 页面使用一个 JavaBean 实例。

（2）< jsp:setProperty >：设置一个 JavaBean 中的属性值。

（3）< jsp:getProperty >：从 JavaBean 中获取一个属性值。

（4）< jsp:include >：在 JSP 页面包含一个外在文件。

（5）< jsp:forward >：把到达的请求转发至另一个页面进行处理。

（6）< jsp:param >：用于传递参数值。

（7）< jsp:plugin >：用于指定在客户浏览器中插入插件的属性。

（8）< jsp:params >：用于向 HTML 页面的插件传递参数值。

（9）< jsp:fallback >：指定如何处理客户端不支持插件运行的情况。

根据各个标准动作的功能，可以将这些标准动作分成以下 5 组。

（1）< jsp:useBean >定义使用一个 JavaBean 实例，id 属性定义了实例名称；< jsp:getProperty >从一个 JavaBean 中获取一个属性值，并将其添加到响应中；< jsp:setProperty >设置一个 JavaBean 中的属性值。

（2）< jsp:include >在请求处理阶段包含来自一个 Servlet 或者 JSP 文件的响应，需要注意与 inducle 指令的不同。

（3）< jsp:forward >将某个请求的处理转发到另一个 Servlet 或者 JSP 页面。

（4）<jsp:param> 用于在其他标准动作中指定参数，对于 <jsp:include> 或 <jsp:forward> 动作还可以将参数传递至另外一个 Servlet 或 JSP 页面。

（5）<jsp:plugin> 的作用是根据浏览器类型在客户端的页面嵌入 Java 对象（例如运行在客户端的 Java Applet 小程序）。

8.5.1 引用外部文件的标准动作 <jsp:include>

该标准动作与前面介绍的 include 指令方法非常类似，也是将特定的外在文件插入当前的页面中，其使用语法如下：

```
<jsp:include page="…url…" flush="true|false"/>
```

该标准动作还可以包含一个体，具体形式如下：

```
<jsp:include page="…url…" flush="true|false"/>
        <jsp:param …/>
</jsp:include>
```

通过在 <jsp:include> 动作体中使用 <jsp:param> 动作，可以用来指定 JSP 页面中可用的其他请求参数，之后可以在当前的 JSP 文件以及引用的外在文件中使用这些请求参数。

专家讲解

@include 指令与 <jsp:include> 标准动作的区别：include 指令是在 JSP 翻译时进行文件的合并，然后对合并的整体文件进行编译；而 <jsp:include> 标准动作则首先进行自身的翻译和编译，然后在用户请求阶段进行二进制文件的合并。

8.5.2 进行请求转移的标准动作 <jsp:forward>

<jsp:forward> 标准动作把请求转移到另外一个页面，这个标准动作只有一个属性 page，page 属性包含一个相对 url 地址，如：

```
<jsp:forward page="/utils/errorReporter.jsp"/>
<jsp:forward page="<%=someJavaExpression %>" />
```

第一行的 page 值是直接给出的，第二行的 <jsp:forward> 标准动作中的 page 值是在请求时动态计算的。

8.5.3 参数设置的标准动作 <jsp:param>

该 <jsp:param> 标准动作一般与 <jsp:include> 以及 <jsp:forward> 等配套使用，用来进行参数的传递，其一般形式如下：

```
<jsp:param name="…名称…" value="…值…"/>
```

每个 <jsp:param> 标准动作都会创建一个既有名又有值的参数，这样就可以使得通过 <jsp:include> 标准动作加载的外在文件以及通过 <jsp:forward> 转发至的另外页面可以使用这些参数，例如：

```
<html>
        <body>
                <jsp:include page="date.jsp">
                        <jsp:param name="serverName" value=" 融创软通 "/>
                </jsp:include>
        </body>
</html>
```

则说明在请求时所包含的 date.jsp 文件可以使用通过 <jsp:param> 标准动作定义的 server-

Name 参数。

8.6 综合案例：使用 <jsp:useBean> 获取表单提交的值

User.java 代码如下。

```
public class User {
    String name;
    String password;
// 省略 getter 和 setter 方法
    }
```

index.jsp 代码如下。

```
<body>
    <jsp:useBean id="user" class="com.isoft.servlet.User"></jsp:useBean>
    <jsp:setProperty property="*" name="user" />
    用户名 :<%=user.getName()%><br>
    密码 :<jsp:getProperty property="password" name="user" />
    <hr>
    <form action="">
            姓名 :<input type="text" name="name" /><br />
密码 :<input type="password" name="password" /><br>
<input type="submit" value=" 提交 " />
    </form>
</body>
```

运行结果如图 8.8 所示。

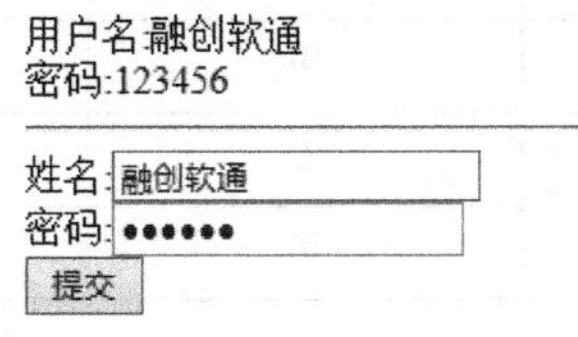

图 8.8　运行结果

小结

JavaBean 的属性的设置方法有如下四种。

（1）自动匹配：< jsp:setProperty name="mybean" property="*"/>。

（2）指定匹配：< jsp:setProperty name="mybean" property="myProperty"/>。

（3）指定传递值和参数的关系：<jsp:setProperty name="mybean" property="myProperty" param="ParamName"/>。

（4）指定值：< jsp:setProperty name="mybean" property="myProperty" value="ParamName"/>。

如果输入的值是变量，需用输出表达式：< jsp:setProperty name="password" property="password" value="<%=password%>"/>。

JavaBean 的属性的取得输出方法为：< jsp:getProperty name="myBean" property="myProperty"/> 值会自动转换类型，将数字的字符串转换成为整型。

经典面试题

1. 什么是 JavaBean？简述 JavaBean 的特点。
2. 一个标准的 JavaBean 如何写？
3. 什么是 JSP 标准动作？
4. 列举出五个以上的标准动作？
5.JSP 中如何将表单输入参数封装到 JavaBean 中？
6.JSP 中动态 include 与静态 include 的区别是什么？
7.JavaBean 中 boolean 属性的 getter 方法的生成规则是什么？

跟我上机

1. 编写一个 Java 类 Student.java 和一个 JSP 页面，把下列信息封装到三个 Student 对象中，再把每个对象存储到一个 ArrayList 对象中，再利用 ArrayList 对象在 JSP 页面的表格中显示所有的信息。

学　号	姓　名	性　别	班　级	成　绩
001	李白	男	01	723.0
002	孟浩然	女	02	543.0
003	白居易	男	03	234.0

2. 编写一个 JSP 程序，做一个 MP3 产品的网上市场调查并提交后，显示出用户填写的市场调查信息。

（1）你是否听说过该品牌？（听说过，没有）

（2）该产品的质量怎么样？（很好，好，一般，差）

（3）……

3. 实现超女音乐吧用户注册功能，使用 JavaBean 和 JSP 标准动作完成如下描述。

（1）注册信息包括用户名、密码、性别、年龄、电话和E-mail，要求：

①用户名不能是Admin；

②用户名、性别、密码和E-mail不能为空；

③密码不低于6位，密码需输入两次，并前后一致；

④ E-mail要求进行合法性验证，建议使用正则表达式。

（2）验证通过后将注册信息显示到本页。

第 9 章　JSP 内置对象

本章要点(学会后请在方框里打钩):

□　了解 JSP 的运行原理

□　掌握 JSP 的九种内置对象

□　掌握 JSP 的属性范围

□　完成在线聊天室案例

JSP 有九个内置对象(又叫隐式对象),是 Web 容器创建的一组对象,不需要预先声明就可以在脚本代码和表达式中随意使用。

9.1 JSP 运行原理

每个 JSP 页面在第一次被访问时,Web 容器都会把请求交给 JSP 引擎(即一个 Java 程序)去处理。JSP 引擎先将 JSP 翻译成一个 _jspServlet(实质上也是一个 Servlet),然后按照 Servlet 的调用方式进行调用。

由于 JSP 第一次访问时会翻译成 Servlet,所以第一次访问通常会比较慢,但第二次访问 JSP 引擎如果发现 JSP 没有变化,就不再翻译,而是直接调用,所以程序的执行效率不会受到影响。

JSP 引擎在调用 JSP 对应的 _jspServlet 时,会传递或创建九个与 Web 开发相关的对象供 _jspServlet 使用。JSP 技术的设计者为便于开发人员在编写 JSP 页面时获得这些 Web 对象的引用,特意定义了九个相应的变量,开发人员在 JSP 页面中通过这些变量就可以快速获得这九个对象的引用。

9.2 认识 JSP 中九个内置对象

JSP 中九个内置对象的类型见表 9.1。

表 9.1 JSP 中九个内置对象类型

序号	内置对象	类 型
1	pageContext	javax.servlet.jsp.PageContext
2	request	javax.servlet.http.HttpServletRequest
3	response	javax.servlet.http.HttpServletResponse
4	session	javax.servlet.http.HttpSession
5	application	javax.servlet.ServletContext
6	config	javax.servlet.ServletConfig
7	out	javax.servlet.jsp.JspWriter
8	page	java.lang.Object
9	exception	java.lang.Throwable

JSP 九个内置对象分为四类,如图 9.1 所示。

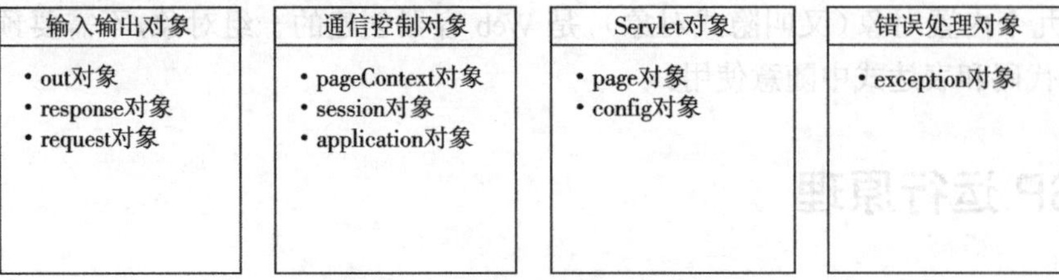

图 9.1 JSP 九个内置对象的分类

（1）pageContext 对象：提供了对 JSP 页面所有对象以及命名空间的访问。

（2）request 对象：封装了来自客户端、浏览器的各种信息。

（3）response 对象：封装了服务器的响应信息。

（4）session 对象：用来保存会话信息，也就是说，可以实现在同一用户的不同请求之间共享数据。

（5）application 对象：代表了当前应用程序的上下文，可以在不同的用户之间共享信息。

（6）config 对象：封装了应用程序的配置信息。

（7）out 对象：用于向客户端、浏览器输出数据。

（8）page 对象：指向了当前 JSP 程序本身。

（9）exception 对象：封装了 JSP 程序执行过程中发生的异常和错误信息。

9.2.1 out 对象

out 对象是一个输出流，用来向浏览器输出信息，除了输出各种信息外还负责对缓冲区进行管理，见表 9.2。

表 9.2 out 对象方法

方法名	说　明
print 或 println	输出数据
newLine	输出换行字符
flush	输出缓冲区数据
close	关闭输出流
clear	清除缓冲区中数据，但不输出到客户端
clearBuffer	清除缓冲区中数据，输出到客户端
getBufferSize	获得缓冲区大小
getRemaining	获得缓冲区中没有被占用的空间
isAutoFlush	是否为自动输出

实例 9.1：out 对象方法的使用

```
<h1>out 内置对象 </h1>
<%
  out.println("<h2> 静夜思 </h2>");// 可以在 println(); 里面加入标签
  out.println(" 床前明月光 <br>");
  out.println(" 疑是地上霜 <br>");
  out.flush();
 // out.clear();// 这里会抛出异常,因为上面有 flush
  out.clearBuffer();// 这里不会抛出异常
  out.println(" 举头望明月 <br>");
  out.println(" 低头思故乡 <br>");
%>
<hr>
缓冲区大小:<%=out.getBufferSize()%>byte<br>
缓冲区剩余大小:<%=out.getRemaining()%>byte<br>
是否自动清空缓冲区:<%=out.isAutoFlush()%><br>
```

运行结果如图 9.2 所示。

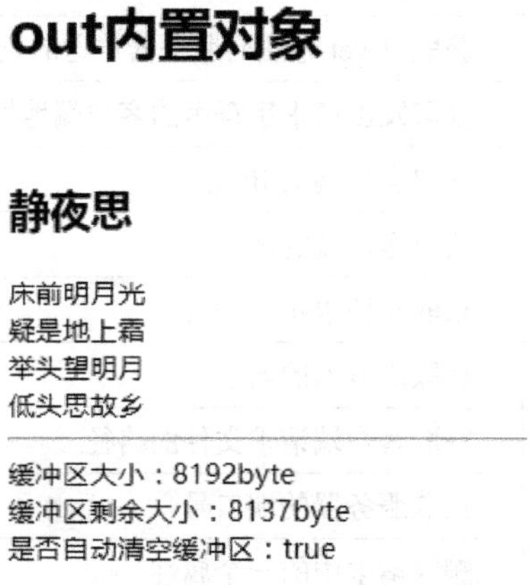

out内置对象

静夜思

床前明月光
疑是地上霜
举头望明月
低头思故乡

缓冲区大小：8192byte
缓冲区剩余大小：8137byte
是否自动清空缓冲区：true

图 9.2　运行结果

9.2.2　request 对象

request 对象封装了从客户端到服务器发出的请求信息，见表 9.3。

表 9.3　request 对象方法

方法名	说　明
isUserInRole	判断认证后的用户是否属于某一成员组
getAttribute	获取指定属性的值，如该属性值不存在返回 null
getAttributeNames	获取所有属性名的集合
getCookies	获取所有 cookie 对象
getCharacterEncoding	获取请求的字符编码方式
getContentLength	返回请求正文的长度，如不确定返回 -1
getHeader	获取指定名字报头值
getHeaders	获取指定名字报头的所有值，一个枚举
getHeaderNames	获取所有报头的名字，一个枚举
getInputStream	返回请求输入流，获取请求中的数据
getMethod	获取客户端向服务器端传送数据的方法
getParameter	获取指定名字参数值
getParameterNames	获取所有参数的名字，一个枚举
getParameterValues	获取指定名字参数的所有值
getProtocol	获取客户端向服务器端传送数据的协议名称
getQueryString	获取以 get 方法向服务器传送的查询字符串
getRequestURI	获取发出请求字符串的客户端地址
getRemoteAddr	获取客户端的 IP 地址
getRemoteHost	获取客户端的名字
getSession	获取和请求相关的会话
getServerName	获取服务器的名字
getServerPath	获取客户端请求文件的路径
getServerPort	获取服务器的端口号
removeAttribute	删除请求中的一个属性
setAttribute	设置指定名字参数值

实例 9.2：完成一个简单的用户注册信息界面，将注册信息发送到欢迎界面上

register.jsp 代码如下。

```
<form action="do_register.jsp" method="post">
  用户名：
  <input type="text" name="userName"><br>
```

```
    技能：
    <input type="checkbox" name="skills" value="Java">Java
    <input type="checkbox" name="skills" value="Oracle">Oracle
    <input type="checkbox" name="skills" value="Spring MVC">Spring MVC
    <input type="checkbox" name="skills" value="JQuery">JQuery
    <br>
    <input type="submit" value=" 提交 ">
    <input type="reset" value=" 重置 ">
</form>
```

do_register.jsp 代码如下。

```
<%
    request.setCharacterEncoding("utf-8");
    // 用户注册信息处理界面使用 getParameter 方法将用户的表单信息提取出来
    String name = request.getParameter("userName");
    String[] skillArr = request.getParameterValues("skills");
// 将 skillArr 数组转换成字符串：
    String skills = "";
    if(skillArr != null && skillArr.length > 0){
        for(String skill : skillArr){
            skills = skills + skill + " ";
        }
    }
// 将数据使用 setAttribute 保存起来
    request.setAttribute("userName", name);
    request.setAttribute("skills", skills);
%>
<%-- 使用 JSP 的 forword 指令将页面跳转到 welcome.jsp--%>
<jsp:forward page="welcome.jsp"></jsp:forward>
```

welcome.jsp 代码如下。

```
信息展示界面：<br><br>
姓名：<%=request.getAttribute("userName")%><br>
技能：<%=request.getAttribute("skills")%>
```

运行结果如图 9.3 所示。

用户名：张建军
技能： Java Oracle Spring MVC JQuery
提交 重置

信息展示界面：

姓名：张建军
技能：Oracle Spring MVC JQuery

图 9.3 运行结果

9.2.2 response 对象

response 对象对客户端的请求做出回应，将 Web 服务器处理后的结果发回客户端。response 对象属于 javax.servlet.HttpServletResponse 接口的实例。response 对象方法见表 9.4。

表 9.4 response 对象方法

方法名	说 明
addCookie	添加一个 cookie 对象
addHeader	添加 HTTP 文件指定名字头信息
containsHeader	判断指定名字 HTTP 文件头信息是否存在
encodeURL	使用 sessionid 封装 URL
flushBuffer	强制把当前缓冲区内容发送到客户端
getBufferSize	返回缓冲区大小
getOutputStream	返回到客户端的输出流对象
sendError	向客户端发送错误信息
sendRedirect	把响应发送到另一个位置进行处理
setContentType	设置响应的 MIME 类型
setHeader	设置指定名字的 HTTP 文件头信息

实例 9.3：response 对象使用

index.jsp 代码如下。

```
<h1>3 s 后跳转到 welcome.jsp 页面，如果没有跳转请按
   <a href="welcome.jsp"> 这里 </a>！ </h1>
<%
   response.setHeader("refresh", "3;URL=welcome.jsp");
%>
```

welcome.jsp 代码如下。

```
<body>
我是自动跳转欢迎页面
</body>
```

运行结果如图 9.4 所示。

图 9.4 运行结果

9.2.4 session 对象

session 对象是一个 JSP 内置对象，它在第一个 JSP 被装载时自动创建，完成会话期管理。从一个客户打开浏览器并连接到服务器开始，到客户关闭浏览器离开这个服务器结束（或者超时），被称为一个会话。当一个客户访问一个服务器时，可能会在这个服务器的几个页面之间切换，服务器需要通过某种办法知道这是一个客户，就需要创建 session 对象。

HTTP 是无状态的连接协议，需要使用 session 来存储用户每次的登录信息。

session 对象方法见表 9.5。

表 9.5 session 对象方法

方法名	说　明
getAttribute	获取指定名字的属性
getAttributeNames	获取 session 中全部属性名字，一个枚举
getCreationTime	返回 session 的创建时间
getId	获取会话标识符
getLastAccessedTime	返回最后发送请求的时间
getMaxInactiveInterval	返回 session 对象的生存时间，单位 0.001 s
invalidate	销毁 session 对象
isNew	每个请求是否会产生新的 session 对象
removeAttribute	删除指定名字的属性
setAttribute	设定指定名字的属性值

实例 9.4：使用 session 对象演示登录应用

login.jsp 代码如下。

```
<body>
欢迎用户登录 <br>
<form action="do_login.jsp" method="post">
  用户名 :<input type="text" name="userName"><br/>
  密    码 :<input type="password" name="password"><br/>
  <input type="submit" value=" 登录 ">
  <input type="reset" value=" 重置 ">
</form>
</body>
```

do_login.jsp 代码如下。

```
<body>
<%
  // 登录逻辑处理界面，使用 getParameter 获取到用户名和密码：
  String userName = request.getParameter("userName");
  String password = request.getParameter("password");
  // 对用户名和密码进行判断：
  if(userName != null && password != null) {
    session.setAttribute("userName", userName);
    response.setHeader("refresh", "2;URL=welcome.jsp");
  }
%>
</body>
```

welcome.jsp 代码如下。

```
<body>
    <%-- 判断 session 对象是否是新创建 --%>
    <%              if(session.isNew()) {   %>
    <br /> 欢迎新用户
    <%              } else { %>
    <br /> 欢迎老用户
    <%              }          %>
    <%              if(session.getAttribute("userName") != null) {    %>
    欢迎
    <%=session.getAttribute("userName")%>
```

```
        <a href="logout.jsp"> 注销 </a>
        <%                  } else { %>
        请先登录
        <a href="Login.jsp"> 登录 </a>
        <%        }          %>
    </body>
```

logout.jsp 代码如下。

```
    <body>
    <% session.invalidate();// 清除掉 session 对象
      response.setHeader("refresh","2;URL=welcome.jsp");
    %>
    </body>
```

运行结果如图 9.5 所示。

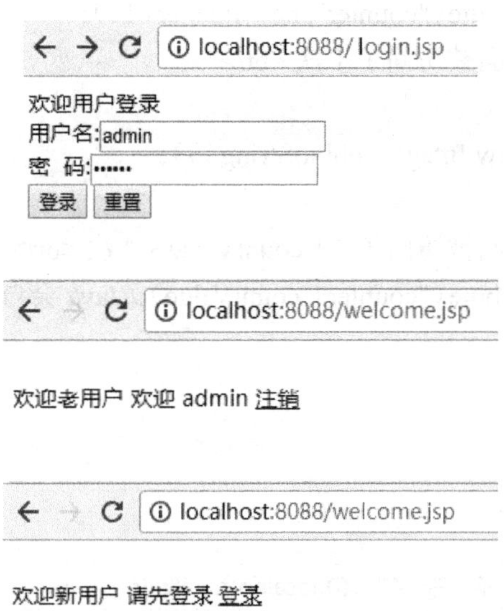

图 9.5 运行结果

9.2.5 application 对象

application 对象代表当前的应用程序，存在于服务器的内存空间中。

应用一旦启动便会自动生成一个 application 对象。如果应用没有被关闭，此 application 对象便会一直存在，直到应用被关闭。application 的生命周期比 session 更长。

application 对象方法见表 9.6。

表 9.6 application 对象方法

方法名	说 明
getAttribute	获取应用对象中指定名字的属性值
getAttributeNames	获取应用对象中所有属性的名字，一个枚举
getInitParameter	返回应用对象中指定名字的初始参数值
getServletInfo	返回 Servlet 编译器中当前版本信息
setAttribute	设置应用对象中指定名字的属性值

实例 9.5：使用 application 对象实现多用户页面访问次数的记录

```
<body>
<%
  Object obj = application.getAttribute("counter");
  if(obj == null) {
    application.setAttribute("counter", new Integer(1));
    out.println(" 该页面被访问了 1 次 <br/>");
  } else {
    int countValue = new Integer(obj.toString());
    countValue++;
    out.println(" 该页面被访问了 " + countValue + " 次 <br/>");
    application.setAttribute("counter", countValue);//Java 会自动装箱
  }
%>
</body>
```

运行结果如图 9.6 所示。

localhost:8088/ login.jsp

该页面被访问了11次

图 9.6 运行结果

9.2.6 config 对象

config 对象表示当前 JSP 程序的配置信息。在一般项目中，JSP 被用作模版技术，也就是位于表示层，而位于表示层的 JSP 文件一般是不需要配置信息的，所以此对象在 JSP 程序中其实很少使用。config 对象是 servletConfig 类的一个实例。config 对象方法见表 9.7。

表 9.7　config 对象方法

方法名	说　明
getServletContext	返回所执行的 Servlet 的环境对象
getServletName	返回所执行的 Servlet 的名字
getInitParameter	返回指定名字的初始参数值
getInitParameterNames	返回该 JSP 中所有的初始参数名，一个枚举

实例 9.6：config 对象使用

web.xml 代码如下。

```
<servlet>
  <servlet-name>config</servlet-name>
  <jsp-file>/Login.jsp</jsp-file>
  <init-param>
    <param-name>name</param-name>
    <param-value> 张建军 </param-value>
  </init-param>
  <init-param>
    <param-name>age</param-name>
    <param-value>30</param-value>
  </init-param>
</servlet>
<servlet-mapping>
  <servlet-name>config</servlet-name>
  <url-pattern>/config</url-pattern>
</servlet-mapping>
```

index.jsp 代码如下。

```
<body>
<!-- 直接输出 config 的 getServletName 的值 -->
<%=config.getServletName()%>
<!-- 输出该 JSP 中名为 name 的参数配置信息 -->
name 配置参数的值:<%=config.getInitParameter("name")%><br/>
<!-- 输出该 JSP 中名为 age 的参数配置信息 -->
age 配置参数的值:<%=config.getInitParameter("age")%>
</body>
```

运行结果如图 9.7 所示。

config
name配置参数的值：张建军
age配置参数的值：30

图 9.7　运行结果

9.2.7　page 对象

page 对象有点类似于 Java 编程中的 this 指针，它指向了当前 JSP 页面本身。

page 对象是 java.lang.object 类的一个实例。

page 对象拥有一个 toString 方法，下面是官方定义的方法介绍：

```
public String toString() {
    return getClass().getName() + "@" + Integer.toHexString(hashCode());
}
// 包名 + 类名 +@+hashcode 值
```

page 对象的方法见表 9.8。

表 9.8　page 对象的方法

方法名	说　明
toString	将当前项目的信息打印出来
getClass	返回当前的 object 类
hashCode	返回 page 对象的 hashCode 值
equals	用于比较对象是否与当前对象相同

实例 9.7：page 对象使用

```
<h1>page 内置对象 </h1>
当前 page 页面对象的字符串描述：<%=page.toString()%><br>
</body>
```

运行结果如图 9.8 所示。

page内置对象

当前page页面对象的字符串描述：org.apache.jsp.Login_jsp@2c8d9c5f

图 9.8　运行结果

9.2.8 exception 对象

exception 对象表示 JSP 引擎在执行代码时抛出的异常。如果想要使用 exception 对象，那么需要配置编译指令的 isErrorPage 属性为 true，即在页面指令中设置 :<%@page isErrorPage="true"%>。

实例 9.8：errorPage 标签使用

login.jsp 代码如下。

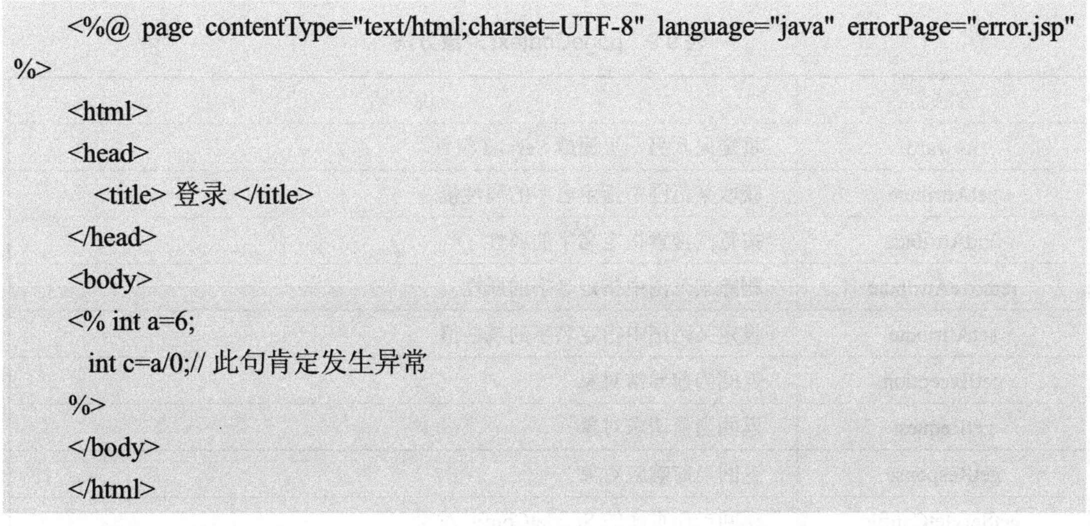

```
<%@ page contentType="text/html;charset=UTF-8" language="java" errorPage="error.jsp" %>
<html>
<head>
  <title> 登录 </title>
</head>
<body>
<% int a=6;
  int c=a/0;// 此句肯定发生异常
%>
</body>
</html>
```

error.jsp 代码如下。

```
<!-- 通过 isErrorPage 属性指定本页面是异常处理页面 -->
<%@ page contentType="text/html;charset=UTF-8" language="java" isErrorPage="true" %>
<html>
<head>
  <title>Title</title>
</head>
<body>
异常类型是 :<%=exception.getClass()%></br>
异常信息是 :<%=exception.getMessage()%></br>
</body>
</html>
```

运行结果如图 9.9 所示。

localhost:8088/login.jsp

异常类型是:class java.lang.ArithmeticException
异常信息是:/ by zero

图 9.9 运行结果

9.2.9 pageContext 对象

pageContext 对象是 javax.servlet.jsp.PageContext 类的实例，主要表示的是一个 JSP 页面的上下文。

pageContetx 对象是 JSP 页面中所有对象功能的最大集成者，使用它可以访问所有的 JSP 内置对象。

pageContext 对象方法见表 9.9。

表 9.9 pageContext 对象方法

方法名	说 明
forward	重定向到另一页面或 Servlet 组件
getAttribute	获取某范围中指定名字的属性值
findAttribute	按范围搜索指定名字的属性
removeAttribute	删除某范围中指定名字的属性
setAttribute	设定某范围中指定名字的属性值
getException	返回当前异常对象
getRequest	返回当前请求对象
getResponse	返回当前响应对象
getServletConfig	返回当前页面的 ServletConfig 对象
getServletContext	返回所有页面共享的 ServletContext 对象
getSession	返回当前页面的会话对象

实例 9.9：pageContext 对象使用

index.jsp 代码如下。

```
<body>
<!-- 括号内？左右不要有空格，否则会报出 HTTP Status 404 -->
<%pageContext.forward("forward.jsp?info= 张建军 ");
%>
</body>
```

forward.jsp 代码如下。

```
<body>
<%
  request.setCharacterEncoding("utf-8");
  // 直接从 pageContext 对象中取得了 request
```

```
    String info = pageContext.getRequest().getParameter("info");
%>
<h3>info=<%=info%></h3>
<h3>realpath=<%=pageContext.getServletContext().getRealPath("/")%></h3>
</body>
```

运行结果如图 9.10 所示。

localhost:8088/index.jsp

info=张建军

realpath=F:\untitled\out\artifacts\untitled_war_exploded

图 9.10 运行结果

9.3 JSP 属性范围

所谓的属性范围就是一个属性设置之后，可以经过多个其他页面后仍然可以访问的保存范围。

JSP 中提供了如下四种属性范围。

（1）当前页：一个属性只能在一个页面中取得，跳转到其他页面无法取得。

（2）一次服务器请求：一个页面中设置的属性，只要经过了服务器跳转，则跳转之后的页面可以继续取得。

（3）一次会话：一个用户设置的内容，只要是与此用户相关的页面都可以访问（一个会话表示一个人，这个人设置的内容只要这个人不走，就依然有效）。

（4）上下文中：在整个服务器上设置的属性，所有人都可以访问。

9.3.1 属性的操作方法

四种属性都包含的属性操作方法见表 9.10。

表 9.10 四种属性都包含的属性操作方法

序 号	方 法	描 述
1	public void setAttribute(String name,Object value)	设置属性
2	public object getAttribute(String name)	取得属性
3	public void removeAttribute(String name)	删除属性

9.3.2 page 属性范围(pageContext)

page 属性范围相对好理解一些:在一个页面设置的属性,跳转到其他页面就无法访问了。但是在使用 page 属性范围的时候必须注意的是,虽然习惯上将页面范围的属性称为 page 范围,但是实际操作的时候是使用 pageContext 内置对象完成的。

pageContext 属性范围操作流程如图 9.11 所示。

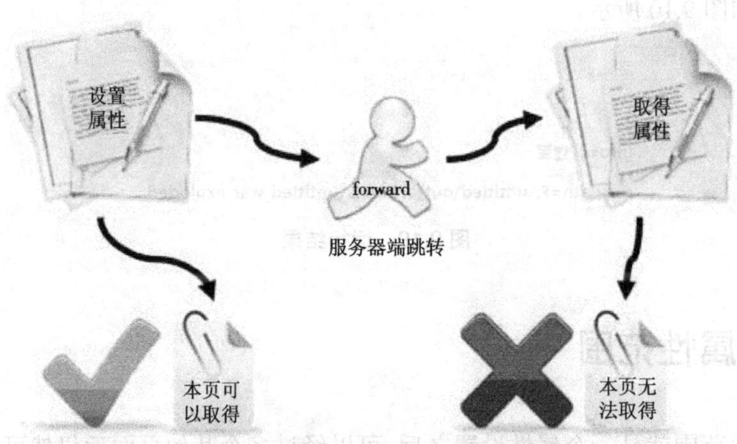

图 9.11 pageContext 属性范围操作流程

从图 9.11 可知,在第一个页面设置的属性经过服务器端跳转到第二个页面以后,在第二个页面是无法获取在第一个页面中设置的属性的,就好比现在坐着的桌子上有一支笔,一旦离开了这张桌子,坐到别的桌子上时,笔就没有了。

实例 9.10:pageContext 对象的使用

index.jsp 代码如下。

```
<body>
<%
    pageContext.setAttribute("name", " 张建军 ");
    pageContext.setAttribute("date", new Date());
%>
<%-- 使用 jsp:forward 标签进行服务器端跳转 --%>
<jsp:forward page="/forward.jsp"/>
</body>
```

forward.jsp 代码如下。

```
<body>
<%
    String refName = (String) pageContext.getAttribute("name");
    Date refDate = (Date) pageContext.getAttribute("date");
```

```
%>
<h1> 姓名:<%=refName%>
</h1>
<h1> 日期:<%=refDate%>
</h1>
</body>
```

运行结果如图 9.12 所示。

localhost:8088/index.jsp

姓名：null

日期：null

图 9.12　运行结果

专家讲解

上面实例使用了服务器端跳转，但是发现内容并不能取得，证明了 page 范围的属性只能在本页中取得，跳转到其他页面之中不能取得。如果现在希望跳转到其他页面之中依然可以取得，则使用 request 属性扩大属性范围即可。

9.3.3　request 属性范围

request 属性范围表示在一次服务器跳转中有效，只要是在服务器中跳转，设置的 request 属性就可以一直传递下去。

request 属性范围操作流程如图 9.13 所示。

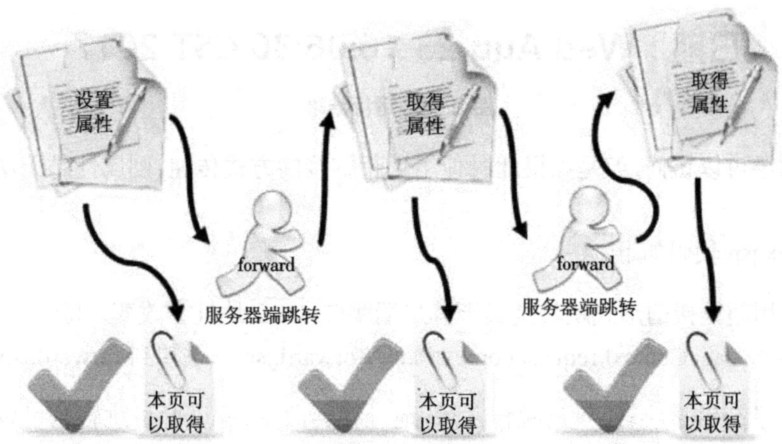

图 9.13　request 属性范围操作流程

实例 9.11：request 对象作用域范围

index.jsp 代码如下。

```
<body>
<%
    request.setAttribute("name", " 张建军 ");
    request.setAttribute("date", new Date());
%>
<%-- 使用 jsp:forward 标签进行服务器端跳转 --%>
<jsp:forward page="/forward.jsp"/>
</body>
```

forward.jsp 代码如下。

```
<%
    String refName = (String) request.getAttribute("name");
    Date refDate = (Date) request.getAttribute("date");
%>
<h1> 姓名：<%=refName%>
</h1>
<h1> 日期：<%=refDate%>
</h1>
```

运行结果如图 9.14 所示。

← → C ⓘ localhost:8088/index.jsp

姓名：张建军

日期：Wed Aug 23 16:06:30 CST 2017

图 9.14　运行结果

以上的结果可以访问，但是如果此时使用了超链接的方式传递，则属性是无法向下继续传递的。

修改 index.jsp 代码如下。

```
<%-- 使用超链接的形式跳转是属于客户端跳转，URL 地址会改变 --%>
<a href="${pageContext.request.contextPath}/forward.jsp"> 跳转到 forward.jsp</a>
```

此时使用了超链接跳转，一旦跳转之后，地址栏会改变，所以此种跳转也可以称为客户端跳转。

forward.jsp 页面显示的结果是 null，这说明了在 index.jsp 这个页面设置的属性经过超链接跳转到别的页面时，别的页面是无法取得 index.jsp 中设置的 request 范围属性的。

如果想取得值应该进一步扩大属性范围，使用 session 范围属性。

9.3.4 session 属性范围

session 设置的属性不管如何跳转，都是可以取得的。当然，session 只针对一个用户。

session 属性范围操作流程如图 9.15 所示。

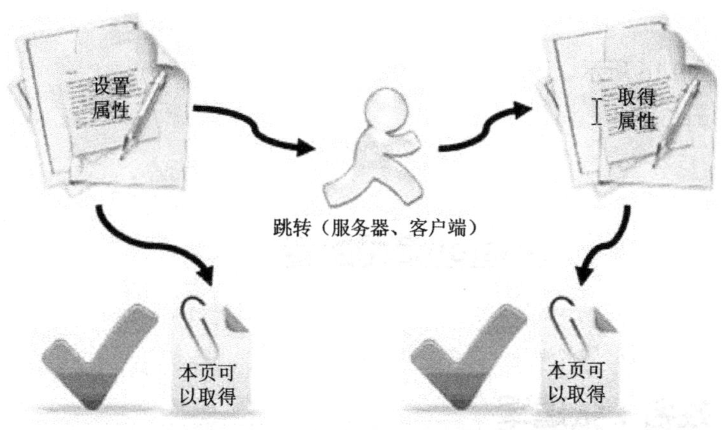

图 9.15 session 属性范围操作流程

在第一个页面上设置的属性，跳转（服务器跳转、客户端跳转）到其他页面之后，其他的页面依然可以取得第一个页面上设置的属性。

实例 9.12：session 对象的使用

index.jsp 代码如下。

```
<body>
<%
    session.setAttribute("name", " 张建军 ");
    session.setAttribute("date", new Date());
%>
<%-- 使用超链接这种客户端跳转 --%>
<h1><a href="${pageContext.request.contextPath}/forward.jsp">session 范围跳转 </a></h1>
</body>
```

forward.jsp 代码如下。

```
<body>
<%
    String refName = (String) session.getAttribute("name");
```

```
    Date refDate = (Date) session.getAttribute("date");
%>
<h1> 姓名:<%=refName%>
</h1>
<h1> 日期:<%=refDate%>
</h1>
</body>
```

运行结果如图 9.16 所示。

← → C ⓘ localhost:8088/index.jsp

session范围跳转

← → C ⓘ localhost:8088/forward.jsp

姓名：张建军

日期：Wed Aug 23 16:19:31 CST 2017

图 9.16　运行结果

专家讲解

以上实例说明了即使是采用客户端跳转，在别的页面依然可以取得第一个页面中设置的 session 属性。但是，如果此时新开了一个浏览器，则 forward.jsp 肯定无法取得 index.jsp 中设置的 session 对象的属性，因为 session 只是保留了一个人的信息。

如果一个属性想让所有的用户都可以访问，则可以使用最后一种属性范围：application 范围。

9.3.5　application 属性范围

application 属性范围操作流程如图 9.17 所示。

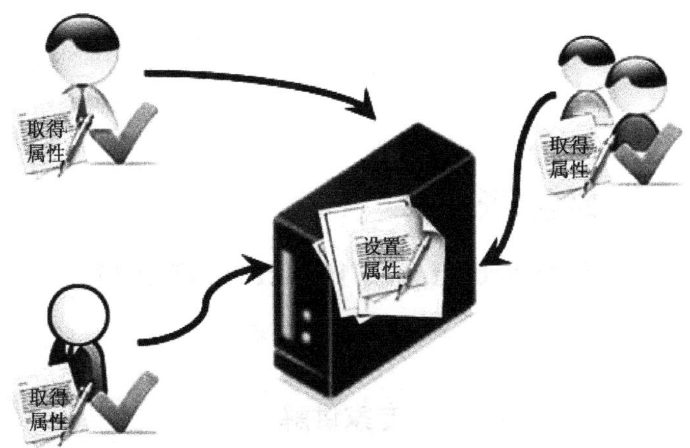

图 9.17 application 属性范围操作流程

因为 application 属性范围是在服务器上设置一个属性，所以一旦设置之后，任何用户都可以浏览到此属性。

实例 9.13：在整个服务器上保存信息，所有用户都可以使用，但重启服务器无法得到属性

index.jsp 代码如下。

```
<%
    application.setAttribute("name", " 张建军 ");
    application.setAttribute("birthday", new Date());
%>
<a href="forward.jsp">application 链接跳转 </a>
```

forward.jsp 代码如下。

```
<body>
<%
    String username = (String) application.getAttribute("name");
    Date birthday = (Date) application.getAttribute("birthday");
%>
<h1> 姓名 :<%=username%>
</h1>
<h2> 出生 :<%=birthday%>
</h2>
</body>
```

运行结果如图 9.18 所示。

图 9.18 运行结果

专家讲解

开启多个浏览器窗口,在运行 forward.jsp 时,都可以显示出图 9.18 所示的结果,因为属性范围设置在服务器中,所以只要是连接到此服务器的任意用户都可以取得此属性,当然,如果服务器关闭,则此属性肯定消失。

注意:如果在服务器上设置了过多的 application 属性,会影响到服务器的性能。

9.3.6 关于 pageContext 属性范围的进一步补充

之前所讲解的四种属性范围,实际上都是通过 pageContext 属性范围设置的。PageContext 类继承了 JspContext 类,在 JspContext 类中定义了 setAttribute 方法,代码如下:

```
public abstract void setAttribute(String name,Object value,int scope)
```

此方法中存在一个 scope 的整型变量,此变量表示一个属性的保存范围。

pageContext 作用域范围如图 9.19 所示。

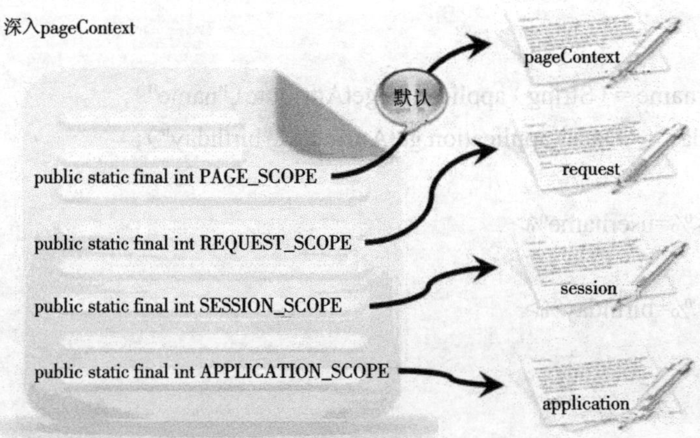

图 9.19 pageContext 作用域范围

pageContext 类继承了 JspContext 类,所以在 PageContext 类中实现了抽象的 setAttribute 方法,代码如下:

```
public abstract void setAttribute(String name,Object value,int scope)
```

这个 setAttribute()方法如果不写后面的 int 类型的 scope 参数，则此参数默认为 PAGE_SCOPE，此时 setAttribute()方法设置的就是 page 属性范围，如果传递过来的 int 类型参数 scope 为 REQUEST_SCOPE，则此时 setAttribute()方法设置的就是 request 属性范围，同理，传递的 scope 参数为 SESSION_SCOPE 和 APPLICATION_SCOPE 时，则表示 setAttribute()方法设置的就是 session 属性范围和 application 属性范围。

9.3.7 JSP 三种属性范围的使用场合

(1)request：如果客户向服务器发请求，产生的数据用户看完就没用，这样的数据就存储在 request 域中，如新闻数据，属于用户看完就没用的。

(2)session：如果客户向服务器发请求，产生的数据用户用完后还有用，像这样的数据就存储在 session 域中，如购物数据，用户需要看到自己购物信息后还要用这个购物数据结账。

(3)application(servletContext)：如果客户向服务器发请求，产生的数据用户用完后还要给其他用户使用，这样的数据就存储在 application(servletContext)域中，如聊天数据。

小结

JSP 隐式对象主要分为四个主要的类别。

(1)输入和输出对象： 控制页面的输入和输出，如 request、response、out。

(2)作用域对象：检索和 JSP 页面的 Servlet 相关的信息，如 pageContext、session、application。

(3)Servlet 对象：提供有关页面环境的信息，如 page、config。

(4)错误对象：处理页面中的错误，如 Exception。

JSP 中有四种范围：page、request、session 和 application，因为这四种属性范围的保存时间不同，所以占用内存的时间也不同，能使用 request 尽量不要使用 session，在一般情况下，request 和 session 两个属性范围使用的频率是最高的。

经典面试题

1. 什么是隐式对象？

2.JSP 的隐式对象有哪些？作用分别是什么？

3.JSP 中有几种属性范围？

4.JSP 隐式对象中，是否有 exception 对象 ?

5. 如何解决提交表单出现乱码的问题?

6. 列举 Request 对象的主要方法。

7.request.getAttribute()和 request.getParameter()有何区别 ?

8. 举例说明页面间对象传递的方法。

9. 详细介绍 session 对象的使用。

10. 详细介绍 request、session、application 三种属性范围的使用场合。

跟我上机

1. 编写一个 JSP 页面，将用户名和密码存储到会话（session）中（假设用户名为“融创软通”，密码为“123456”），再重新定向到另一个 JSP 页面，将会话中存储的用户名和密码显示出来。

（提示：①使用 response 对象的 sendRedirect（）方法进行重定向；②使用 session 的 setAttribute 和 getAttribute 方法。）

2. 编写一格 JSP 登录页面，可输入用户名和密码，提交请求到另一个 JSP 页面，在该 JSP 页面获取请求的相关数据并显示出来。请求的相关数据包括用户输入的请求数据和请求本身的一些信息（如请求使用的协议、请求的 URL 等）。

3. 编写一个 JSP 页面，产生随机数作为用户幸运数字（1 位），将其保存到会话中，并重新定向到另一个页面，将用户的幸运数字显示出来。

4. 利用隐式对象为某一网站编写一个 JSP 程序，统计该网页的运行次数。

（提示：用户每打开一次窗口运行该网页或在同一窗口刷新该网页都算运行一次，可利用 application 对象去实现。）

完成功能如下图所示。

计算结果:43+4=47.0

简单的计数器

第一个参数

操作符 +

第二个参数

计算

第 10 章　EL 表达式

EL表达式

本章要点(学会后请在方框里打钩):

- □ 了解什么是 EL 表达式
- □ 掌握 EL 表达式的运算符
- □ 掌握 EL 表达式中内置对象的使用
- □ 掌握 EL 表达式函数库的使用
- □ 了解 EL 表达式的保留字

10.1 EL 表达式简介

EL（表达式语言，Expression Language）的目的是为了使 JSP 写起来更加简单。

表达式语言的灵感来自于 ECMAScript 和 XPath 表达式语言，它提供了在 JSP 中简化表达式的方法。它是一种简单的语言，基于可用的命名空间（PageContext 属性）、嵌套属性和对集合、操作符（算术型、关系型和逻辑型）的访问符、映射到 Java 类中静态方法的可扩展函数以及一组隐式对象。

EL 提供了在 JSP 脚本编制元素范围外使用运行时表达式的功能。脚本编制元素是指页面中能够用于在 JSP 文件中嵌入 Java 代码的元素。它们通常用于对象操作以及执行那些影响所生成内容的计算。JSP 2.0 将 EL 表达式添加为一种脚本编制元素。

10.1.1 获取数据

使用 EL 表达式获取数据语法：

```
"${ 标识符 }"
```

EL 表达式语句在执行时，会调用 pageContext.findAttribute 方法，用标识符为关键字，分别从 page、request、session、application 四个域中查找相应的对象，找到则返回相应对象，找不到则返回空字符串。

EL 表达式可以很轻松地获取 JavaBean 的属性，或获取数组、Collection、Map 类型集合的数据。

实例 10.1：EL 表达式获取数据实例

person.java 代码如下。

```
public class Person {
   //Person 类封装的私有属性
   private String name;
   private String sex;
   private int age;
   private boolean married;
   private Date birthday;
   private  Address address;
 // 省略 getter 和 setter 方法
 }
```

address.java 代码如下。

```
public class Address {
  private  String name;
```

```
// 省略 getter 和 setter 方法
}
```

index.jsp 代码如下。

```
<%@ page language="java" import="java.util.*" pageEncoding="UTF-8" %>
<%@taglib uri="http://java.sun.com/jsp/jstl/core" prefix="c" %>
<%@page import="com.iss.bean.Person" %>
<%@page import="com.iss.bean.Address" %>
<!DOCTYPE HTML>
<html>
<head>
  <title>EL 表达式获取数据 </title>
</head>
<body>
<%
  request.setAttribute("name", " 张建军 ");
%>
<%--${name} 等同于 pageContext.findAttribute("name") --%>
使用 EL 表达式获取数据:${name}
<hr>
<!-- 在 JSP 页面中,使用 EL 表达式可以获取 bean 的属性 -->
<%
  Person p = new Person();
  p.setAge(12);
  request.setAttribute("person", p);
%>   使用 EL 表达式可以获取 bean 的属性:${person.age}
<hr>
<!-- 在 JSP 页面中,使用 EL 表达式可以获取 bean 中的属性 -->
<%
  Person person = new Person();
  Address address = new Address();
  person.setAddress(address);
  request.setAttribute("person", person);
%>
${person.address.name}
<hr>
<!-- 在 JSP 页面中,使用 EL 表达式获取 list 集合中指定位置的数据 -->
<%
```

```
Person p1 = new Person();
  p1.setName(" 张建军 ");
  Person p2 = new Person();
  p2.setName(" 何晶 ");
  List<Person> list = new ArrayList<Person>();
  list.add(p1);
  list.add(p2);
  request.setAttribute("list", list);
%>
<!-- 取 list 指定位置的数据 -->
${list[1].name}
<!-- 迭代 List 集合 -->
<c:forEach var="person" items="${list}">
  ${person.name}
</c:forEach>
<hr>
<!-- 在 JSP 页面中，使用 EL 表达式获取 map 集合的数据 -->
<%
  Map<String, String> map = new LinkedHashMap<String, String>();
  map.put("a", "AAAAA");
  map.put("b", "BBBBB");
  map.put("c", "CCCCC");
  map.put("d", "DDDDD");
  request.setAttribute("map", map);
%>
<!-- 根据关键字取 map 集合的数据 -->
${map.c}
${map["d"]}
<hr>
<!-- 迭代 map 集合 -->
<c:forEach var="me" items="${map}">
  ${me.key}=${me.value}<br/>
</c:forEach>
<hr>
</body>
</html>
```

运行结果如图 10.1 所示。

localhost:8088/index.jsp

使用EL表达式获取数据：张建军

使用EL表达式可以获取bean的属性：12

何晶 张建军 何晶

CCCCC DDDDD

a=AAAAA
b=BBBBB
c=CCCCC
d=DDDDD

图 10.1 运行结果

10.1.2 执行运算

执行运算语法：

```
${ 运算表达式 }
```

EL 表达式支持表 10.1 所示运算符。

表 10.1 EL 表达式支持的运算符

运算符分类	定 义
算术运算符	+、-（二元）、*、/、div、%、mod、-（一元）
逻辑运算符	and、&&、or、||、!、not
关系运算符	==、eq、!=、ne、gt、<=、le、>=、ge。可以与其他值进行比较，或与布尔型、字符串型、整型或浮点型文字进行比较
空运算符	空操作符是前缀操作，可用于确定值是否为空
条件运算符	A ?B :C。根据 A 赋值的结果来赋值 B 或 C

实例 10.2：EL 表达式支持的运算符

User.java 代码如下。

```
public class User {
   private String  username;
   private  String gender;
// 省略 setter 和 getter
}
```

index.jsp 代码如下。

```
<%@ page language="java" import="java.util.*" pageEncoding="UTF-8" %>
<%@taglib uri="http://java.sun.com/jsp/jstl/core" prefix="c" %>
<%@page import="com.iss.bean.User" %>
```

```
<!DOCTYPE HTML>
<html>
<head>
  <title>EL 表达式计算数据 </title>
</head>
<body>
<h3>EL 表达式进行四则运算:</h3>
加法运算:${365+24}<br/>
减法运算:${365-24}<br/>
乘法运算:${365*24}<br/>
除法运算:${365/24}<br/>

<h3>EL 表达式进行关系运算:</h3>
<%--${user == null} 和 ${user eq null} 两种写法等价 --%>
${user == null}<br/>
${user eq null}<br/>

<h3>EL 表达式使用 empty 运算符检查对象是否为 null(空)</h3>
<%
  List<String> list = new ArrayList<String>();
  list.add(" 张建军 ");
  list.add(" 何晶 ");
  request.setAttribute("list", list);
%>
<%-- 使用 empty 运算符检查对象是否为 null(空) --%>
<c:if test="${!empty(list)}">
  <c:forEach var="str" items="${list}">
    ${str}<br/>
  </c:forEach>
</c:if>
<br/>
<%
  List<String> emptyList = null;
%>
<%-- 使用 empty 运算符检查对象是否为 null(空) --%>
<c:if test="${empty(emptyList)}">
  对不起,没有您想看的数据
</c:if>
```

```
<br/>
<h3>EL 表达式中使用二元表达式 </h3>
<%
  session.setAttribute("user", new User(" 张建军 "));
%>
${user==null? " 对不起，您没有登录 " : user.username}

<br/>
<h3>EL 表达式数据回显 </h3>
<%
  User user = new User(" 张建军 ");
  user.setGender(" 男 ");
  // 数据回显
  request.setAttribute("user", user);
%>
<input type="radio" name="gender" value=" 男 " ${user.gender==' 男 '?'checked':''}> 男
<input type="radio" name="gender" value=" 女 " ${user.gender==' 女 '?'checked':''}> 女
<br/>
</body>
</html>
```

运行结果如图 10.2 所示。

```
localhost:8088/index.jsp

EL表达式进行四则运算：
加法运算：389
减法运算：341
乘法运算：8760
除法运算：15.208333333333334

EL表达式进行关系运算：
true
true

EL表达式使用empty运算符检查对象是否为null(空)
张建军
何晶

对不起，没有您想看的数据

EL表达式中使用二元表达式
张建军

EL表达式数据回显
男 女
```

图 10.2　运行结果

10.1.3 EL 表达式内置对象

EL 表达式语言中定义了 11 个隐含对象，使用这些隐含对象可以很方便地获取 Web 开发中的一些常见对象，并读取这些对象的数据。语法：

```
${ 隐式对象名称 }：获得对象的引用
```

EL 表达式的内置对象见表 10.2。

表 10.2 EL 表达式的内置对象

序号	隐含对象名称	描述
1	pageContext	对应于 JSP 页面中的 pageContext 对象（注意：取的是 pageContext 对象）
2	pageScope	代表 page 域中用于保存属性的 Map 对象
3	requestScope	代表 request 域中用于保存属性的 Map 对象
4	sessionScope	代表 session 域中用于保存属性的 Map 对象
5	applicationScope	代表 application 域中用于保存属性的 Map 对象
6	param	表示一个保存了所有请求参数的 Map 对象
7	paramValues	表示一个保存了所有请求参数的 Map 对象，它对于某个请求参数，返回的是一个 string[]
8	header	表示一个保存了所有 HTTP 请求头字段的 Map 对象，注意：如果头里面有“-”，例 Accept-Encoding，则要 header["Accept-Encoding"]
9	headerValues	表示一个保存了所有 HTTP 请求头字段的 Map 对象，它对于某个请求参数，返回的是一个 string[] 数组。注意：如果头里面有“-”，例 Accept-Encoding，则要 headerValues["Accept-Encoding"]
10	cookie	表示一个保存了所有 cookie 的 Map 对象
11	initParam	表示一个保存了所有 Web 应用初始化参数的 Map 对象

测试 EL 表达式中的 11 个隐式对象代码如下。

```
<%@ page language="java" import="java.util.*" pageEncoding="UTF-8" %>
<!DOCTYPE HTML>
<html>
<head>
  <title>EL 隐式对象 </title>
</head>
<body>
<br/>1.pageContext 对象：获取 JSP 页面中的 pageContext 对象 <br/>
```

```
${pageContext}
<br/>2.pageScope 对象：从 page 域（pageScope）中查找数据 <br/>
<%
  pageContext.setAttributc（"namc", " 张建军 "）; //map
%>
${pageScope.name}
<br/>3.requestScope 对象：从 request 域（requestScope）中获取数据 <br/>
<%
  request.setAttribute（"name", " 何晶 "）; //map
%>   ${requestScope.name}
<br/>4.sessionScope 对象：从 session 域（sessionScope）中获取数据 <br/>
<%
  session.setAttribute（"user", " 融创软通 "）; //map
%>
${sessionScope.user}
<br/>5.applicationScope 对象：从 application 域（applicationScope）中获取数据 <br/>
<%
  application.setAttribute（"user", " 高雅 "）; //map
%>
${applicationScope.user}
<br/>6.param 对象：获得用于保存请求参数 map，并从 map 中获取数据 <br/>
${param.name}
<!-- 此表达式会经常用在数据回显上 -->
<form action="RegisterServlet" method="post">
  <input type="text" name="username" value="${param.username}">
  <input type="submit" value=" 注册 ">
</form>
<br/>7.paramValues 对象：paramValues 获得请求参数 //map{"",String[]}<br/>
<!-- http://localhost:8080/index.jsp?like=aaa&like=bbb -->
${paramValues.like[0]}
${paramValues.like[1]}
<br/>8.header 对象：header 获得请求头 <br/>
${header.Accept}<br/>
${header["Accept-Encoding"]}
<br/>9.headerValues 对象：headerValues 获得请求头的值 <br/>
<%--headerValues 表示一个保存了所有 HTTP 请求头字段的 map 对象，它对于某个请求
参数，返回的是一个 string[] 数组
```

```
例如：headerValues.Accept 返回的是一个 string[] 数组，headerValues.Accept[0] 取出数组中的第一个值
--%>
${headerValues.Accept[0]}<br/>
<%--${headerValues.Accept-Encoding} 这样写会报错
  测试 headerValues 时，如果头里面有"-"，例 Accept-Encoding，则要 headerValues["Accept-Encoding"]
     headerValues["Accept-Encoding"] 返回的是一个 string[] 数组，headerValues["Accept-Encoding"][0] 取出数组中的第一个值
--%>
${headerValues["Accept-Encoding"][0]}
<br/>10.cookie 对象：cookie 对象获取客户机提交的 cookie<br/>
<!-- 从 cookie 隐式对象中根据名称获取到的是 cookie 对象，要想获取值，还需要 .value -->
${cookie.JSESSIONID.value} // 保存所有 cookie 的 Map
<br/>11.initParam 对象：initParam 对象获取在 web.xml 文件中配置的初始化参数 <br/>
<%--
<!-- web.xml 文件中配置初始化参数 -->
<context-param>
  <param-name>xxx</param-name>
  <param-value>yyyy</param-value>
</context-param>    <context-param>
 <param-name>root</param-name>
  <param-value>/config</param-value>
</context-param>
--%>
<%-- 获取 ServletContext 中用于保存初始化参数的 map --%>
${initParam.xxx}<br/>
${initParam.root}
</body>
</html>
```

RegisterServlet 代码如下。

```
@WebServlet(name = "RegisterServlet ")
public class RegisterServlet extends HttpServlet {
  protected void doPost(HttpServletRequest request, HttpServletResponse response) throws ServletException, IOException {
```

```
        String userName = request.getParameter("username");
        request.getRequestDispatcher("/index.jsp").forward(request, response);
    }
}
```

测试结果如图 10.3 所示。

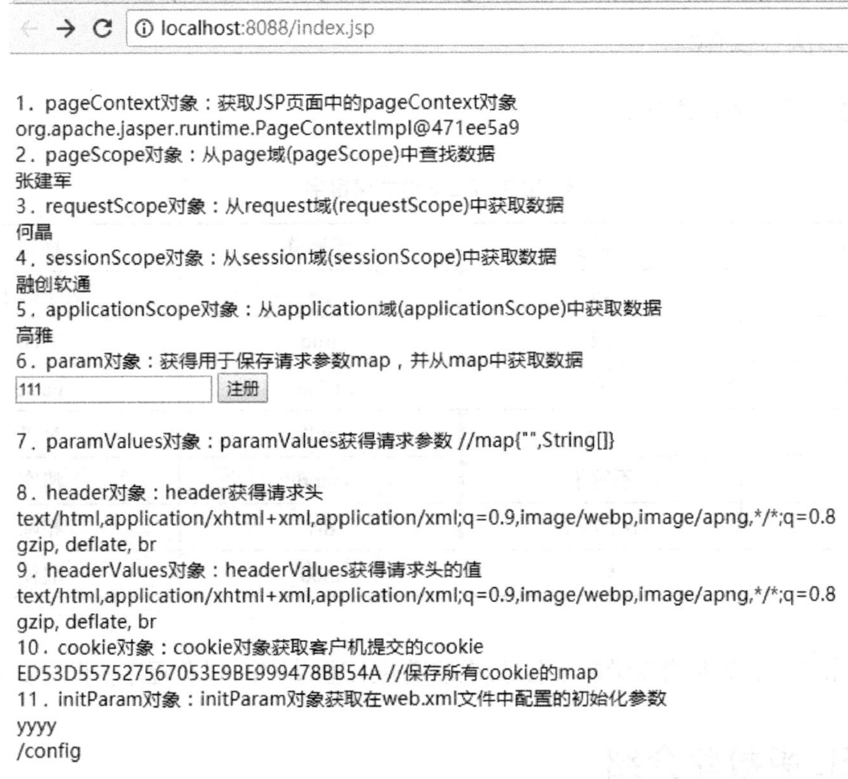

图 10.3　测试结果

10.1.4　使用 EL 调用 Java 方法

EL 表达式语法允许开发人员开发自定义函数，以调用 Java 类的方法。语法：

```
${prefix:method(params)}
```

在 EL 表达式中调用的只能是 Java 类的静态方法，这个 Java 类的静态方法需要在 TLD 文件中描述，才可以被 EL 表达式调用。

EL 自定义函数用于扩展 EL 表达式的功能，可以让 EL 表达式完成普通 Java 程序代码所能完成的功能。

本节只需了解即可，用得不多，不做太多要求。

10.1.5 EL 注意事项

EL 表达式是 JSP 2.0 规范中的一门技术。因此，若想正确解析 EL 表达式，需使用支持 Servlet 2.4/JSP 2.0 技术的 Web 服务器。

注意：有些 Tomcat 服务器如不能使用 EL 表达式，需要提前完成如下设置。

（1）升级到 Tomcat 6 以上

（2）在 JSP 中加入 <%@ page isELIgnored="false" %>。

10.1.6 EL 表达式保留字

EL 表达式保留字见表 10.3。

表 10.3 EL 表达式保留字

运算符	描述	运算符	描述
and	与	ge	大于等于
or	或	true	True
not	非	false	False
eq	等于	null	Null
ne	不等于	empty	清空
le	小于等于	div	相除
gt	大于	mod	取模

所谓保留字的意思是指变量在命名时，应该避开上述的名字，以免程序编译时发生错误。

10.2 EL 函数库介绍

由于在 JSP 页面中显示数据时，经常需要对显示的字符串进行处理，SUN 公司针对一些常见处理定义了一套 EL 函数库供开发者使用。

EL 表达式函数库见表 10.4。

表 10.4 EL 表达式函数库

函数名	函数说明	使用举例
fn:contains	判断字符串是否包含另外一个字符串	<c:if test="${fn:contains(name, searchString)}">
fn:containsIgnore-Case	判断字符串是否包含另外一个字符串（大小写无关）	<c:if test="${fn:containsIgnoreCase(name, searchString)}">

续表

函数名	函数说明	使用举例
fn:endsWith	判断字符串是否以另外字符串结束	<c:if test="${fn:endsWith(filename, ".txt")}">
fn:escapeXml	把一些字符转成 XML 表示,例如 < 字符应该转为 <	${fn:escapeXml(param:info)}
fn:indexOf	子字符串在母字符串中出现的位置	${fn:indexOf(name, "-")}
fn:join	将数组中的数据联合成一个新字符串,并使用指定字符格开	${fn:join(array, ";")}
fn:length	获取字符串的长度,或者数组的大小	${fn:length(shoppingCart.products)}
fn:replace	替换字符串中指定的字符	${fn:replace(text, "-", "•")}
fn:split	把字符串按照指定字符切分	${fn:split(customerNames, ";")}
fn:startsWith	判断字符串是否以某个子串开始	<c:if test="${fn:startsWith(product.id, "100-")}">
fn:substring	获取子串	${fn:substring(zip, 6, -1)}
fn:substringAfter	获取从某个字符所在位置开始的子串	${fn:substringAfter(zip, "-")}
fn:substringBefore	获取从开始到某个字符所在位置的子串	${fn:substringBefore(zip, "-")}
fn:toLowerCase	转为小写	${fn.toLowerCase(product.name)}
fn:toUpperCase	转为大写字符	${fn.UpperCase(product.name)}
fn:trim	去除字符串前后的空格	${fn.trim(name)}

这些 EL 函数在 JSTL 开发包中进行描述,因此在 JSP 页面中使用 SUN 公司的 EL 函数库,需要导入 JSTL 开发包,并在页面中导入 EL 函数库。

在页面中使用 JSTL 定义的 EL 函数代码如下:

```
<%@taglib uri="http://java.sun.com/jsp/jstl/functions" prefix="fn"%>
```

10.3 综合案例:使用 EL 函数库中的方法

```
<%@ page language="java" import="java.util.*" pageEncoding="UTF-8" %>
<%@taglib prefix="fn" uri="http://java.sun.com/jsp/jstl/functions" %>
<%@page import="com.isoft.bean.User" %>
<!DOCTYPE HTML>
<html>
<head>
```

```
    <title>EL 函数库中的方法使用案例 </title>
  </head>
  <body>
  <h3>fn:toLowerCase 函数使用案例:</h3>
  fn:toLowerCase("WWW.91isoft.com") 的 结 果 是: ${fn:toLowerCase("WWW.91isoft.
com")}
  <hr/>
  <h3>fn:toUpperCase 函数使用案例:</h3>
  fn:toUpperCase("WWW.91isoft.com") 的 结 果 是: ${fn:toUpperCase("WWW.91isoft.
com")}
  <hr/>
  <h3>fn:trim 函数使用案例:</h3>
  fn:trim(" www.91isoft.com ")的结果是:${fn:trim(" www.91isoft.com ")}
  <hr/>
  <h3>fn:length 函数使用案例:</h3>
  <%
    List<String> list = Arrays.asList("1", "2", "3");
    request.setAttribute("list", list);
  %>
  fn:length(list)计算集合 list 的 size 的值是:${fn:length(list)}
  <br/>
  fn:length("www.91isoft.com")计算字符串的长度是:${fn:length("www.91isoft.com")}
  <hr/>
  <h3>fn:split 函数使用案例:</h3>
  fn:split("www.91isoft.com",".")[0] 的结果是:${fn:split("www.91isoft.com",".")[0]}
  <hr/>
  <h3>fn:join 函数使用案例:</h3>
  <%
    String[] StringArray = {"www", "91isoft", "com"};
    pageContext.setAttribute("StringArray", StringArray);
  %>
  fn:join(StringArray,".")的结果是:${fn:join(StringArray,".")}
  <br/>
  fn:join(fn:split("www,91isoft,com",","),".") 的 结 果 是: ${fn:join(fn:split("www,91isoft-
,com",","),".")}
  <hr/>
  <h3>fn:indexOf 函数使用案例:</h3>
```

```
    fn:indexOf("www.91isoft.com","isoft")的返回值为：${fn:indexOf("www.91isoft.com","i-
soft")}
    <hr/>
    <h3>fn:contains 函数使用案例：</h3>
    <%
      User user = new User();
      String likes[] = {"sing", "dance"};
      user.setLikes(likes);
      // 数据回显
      request.setAttribute("user", user);
    %>
    <%-- 使用 EL 函数回显数据 --%>
    <input type="checkbox" name="like"
        vlaue="sing" ${fn:contains(fn:join(user.likes,","),"sing")?'checked':''}/> 唱歌
    <input type="checkbox" name="like"
        value="dance"  ${fn:contains(fn:join(user.likes,","),"dance")?'checked':''}/> 跳舞
    <input type="checkbox" name="like"
          value="basketball" ${fn:contains(fn:join(user.likes,","),"basketball")?'checked':''}/>
蓝球
    <input type="checkbox" name="like"
        value="football"  ${fn:contains(fn:join(user.likes,","),"football")?'checked':''}/> 足球
    <hr/>
    <h3>fn:startsWith 函数和 fn:endsWith 函数使用案例：</h3>
    fn:startsWith("www.91isoft.com","91isoft") 的返回值为：${fn:startsWith("www.91isoft.
com","91isoft")}
    <br/>
    fn:endsWith("www.91isoft.com","com") 的返回值为：${fn:endsWith("www.91isoft.
com","com")}
    <hr/>
    <h3>fn:replace 使用案例：</h3>
    fn:replace("www 91isoft com ", " ", ".")的返回值为字符串：${fn:replace("www 91isoft
com", " ", ".")}
    <hr/>
    <h3>fn:substring 使用案例：</h3>
    fn:substring("www.91isoft.org", 4, 9) 的返回值为字符串：${fn:substring("www.91isoft.
org", 4, 9)}
    <h3>fn:substringAfter 函数和 fn:substringBefore 函数使用案例：</h3>
```

```
fn:substringAfter("www.91isoft.org",".")的返回值为字符串：${fn:substringAfter("www.91isoft.org",".")}
<br/>
fn:substringBefore("www.91isoft.org",".")的返回值为字符串：${fn:substringBefore("www.91isoft.org",".")}
<hr/>
</body>
</html>
```

运行结果如下。

fn:toLowerCase 函数使用案例：

fn:toLowerCase("WWW.91isoft.com")的结果是：www.91isoft.com

---

fn:toUpperCase 函数使用案例：

fn:toUpperCase("WWW.91isoft.com")的结果是：WWW.91ISOFT.COM

---

fn:trim 函数使用案例：

fn:trim(" www.91isoft.com ")的结果是：www.91isoft.com

---

fn:length 函数使用案例：

fn:length(list)计算集合 list 的 size 的值是：3

fn:length("www.91isoft.com")计算字符串的长度是：15

---

fn:split 函数使用案例：

fn:split("www.91isoft.com",".")[0] 的结果是：www

---

fn:join 函数使用案例：

fn:join(StringArray,".")的结果是：www.91isoft.com

fn:join(fn:split("www,91isoft,com",","),".")的结果是：www.91isoft.com

---

fn:indexOf 函数使用案例：

fn:indexOf("www.91isoft.com","isoft")的返回值为：6

---

fn:contains 函数使用案例：

☑ 唱歌 ☑ 跳舞 ☐ 蓝球 ☐ 足球

---

fn:startsWith 函数和 fn:endsWith 函数使用案例：

fn:startsWith("www.91isoft.com","91isoft")的返回值为：false

fn:endsWith("www.91isoft.com","com")的返回值为:true

---

fn:replace 使用案例:

fn:replace("www 91isoft com ", " ", ".")的返回值为字符串:www.91isoft.com

---

fn:substring 使用案例:

fn:substring("www.91isoft.org", 4, 9)的返回值为字符串:91iso

fn:substringAfter 函数和 fn:substringBefore 函数使用案例:

fn:substringAfter("www.91isoft.org",".")的返回值为字符串:91isoft.org

fn:substringBefore("www.91isoft.org",".")的返回值为字符串:www

小结

通过 EL 可以简化在 JSP 开发中对对象的引用,从而规范页面代码,增加程序的可读性及可维护性。

EL 表达式的语法有两个要素:$ 和 {}。

EL 表达式可以使用"."或者"[]"操作符在相应的作用域中取得某个属性的值。

EL 表达式的几个特点如下。

(1)可以与 JSP 标签库结合使用,也可以与 JavaScript 语句集合使用。

(2)EL 会自动进行类型转换。如果想求两个字符串型数值的和,可以直接通过"+"进行连接,比如:${num1+num2}。

(3)在 EL 中可以获得命名空间(PageContext 对象是页面中所有内置对象的最大范围的集成对象,通过它可以访问其他内置对象)。

(4)在 EL 表达式中可以访问 JSP 的作用域(request,application,page,session)。

经典面试题

1. 介绍一下 EL 表达式语言。
2. 列举一下 EL 表达式的四种范围对象。
3.JSP 脚本和 EL 表达式的区别。
4.JSP EL 表达式中 float 型转成 int 型应如何做?
5. 如何禁用 JSP 中的 EL 表达式?
6. 如何用 EL 表达式在 JSP 中获得系统当前时间?
7.EL 表达式中 list 有取长度的方法吗?
8.EL 表达式如何截取字符串中的部分数据?
9. 为什么需要 EL?
10. 简要说明 EL 表达式中的隐式对象。

跟我上机

1. 用表达式语言 EL 算术运算符编写一个 JSP 页面程序，以生成一个账单，该账单应显示产品编号、产品名称、成本价格、数量和金额，还应显示合计金额。

2. 编写一个 JSP 程序，用以获取一名学员五门学科的成绩，每门学科的得分均在 100 分以内，在同一个页面上显示所提交分数的总分和平均分，如下图所示。

五门学科的平均分

输入英语分数：4 /100
输入化学分数：4 /100
输入历史分数：4 /100
输入地理分数：4 /100
输入数学分数：4 /100
总分： 20 /500
平均分： 4.0
提交

3. 编写一个 JSP 页面，使用 JSP 表达式（EL）语言，显示九九乘法表。

4. 使用 JSP 表达式的关系运算符，比较两个数值的大小。

5. 使用 JSP 表达式语言配置 Context 和 Scope 对象，改变窗体的背景颜色和文字的大小。

6. 使用 EL 表达式逻辑运算符完成下图功能。

逻辑 EL

运算	EL 表达式	结果
与	${true and true}	true
与	${true && false}	false
或	${true or true}	true
或	${true || false}	true
非	${not true}	false
非	${!false}	true

7. 从数据库中读取一些记录，制作成从下向上流动的跑马灯动画效果。（做成相应的超链接）

8. 创建一个基于 JSP 的应用，取 1~10 作为应用，然后打印出一个简单的乘法表。例如，如果输入 5，显示的结果是 5 time 1 is 5,5 time 2 is 10,5 time 3 is 15……

9. 使用 EL 表达式完成下图所示功能。

10. 使用 EL 表达式完成下图所示功能。

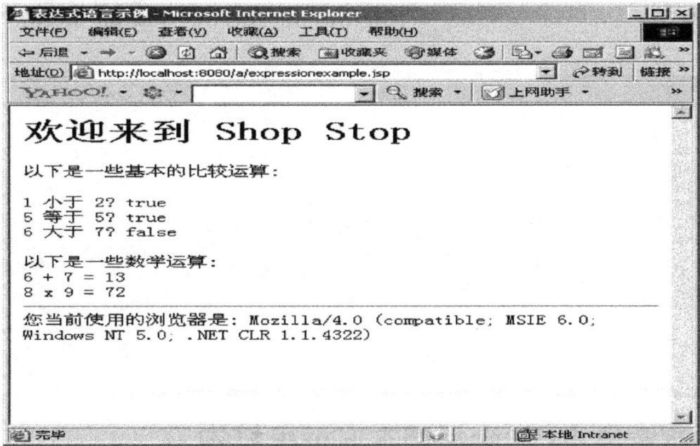

第 11 章　JSTL 标准标签库

JSTL Introducing The JSP Standard Tag Library Shawn Bayern

本章要点(学会后请在方框里打钩):

- □ 了解什么是 JSTL
- □ 掌握 JSTL 的分类
- □ 熟练掌握 JSTL 中核心标签库的使用
- □ 熟练掌握 JSTL 中流程控制标签的使用
- □ 掌握 JSTL 中格式化标签的使用

11.1 JSTL 标签库介绍

JSTL 全名为 JavaServer Pages Standard Tag Library，JSTL 标签库是为弥补 HTML 标签的不足，规范自定义标签的使用而诞生的。使用 JSTL 标签的目的就是不希望在 JSP 页面中出现 Java 逻辑代码。

JSTL 是由 JCP（Java Community Process）所制定的标准规范，它主要提供给 Java Web 开发人员一个标准通用的标签函数库。

Web 程序员能够利用 JSTL 和 EL 来开发 Web 程序，取代传统直接在页面上嵌入 Java 程序（Scripting）的做法，以提高程序的阅读性、维护性和方便性。

11.2 JSTL 标签库的分类

JSTL 标签库共分为五大类，如图 11.1 所示。

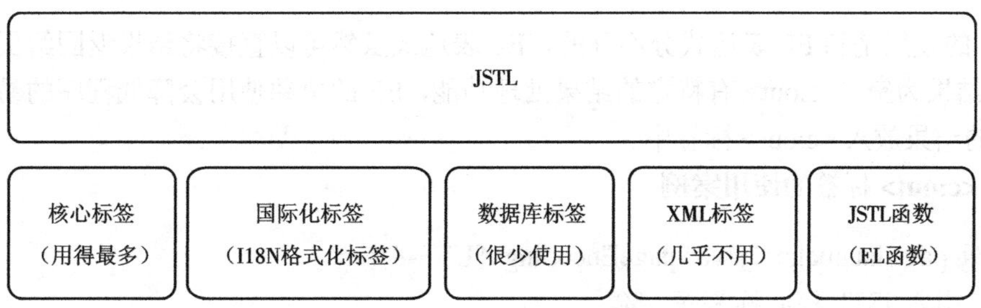

图 11.1 JSTL 标签库分类

11.3 核心标签库使用说明

JSTL 的核心标签库标签共有 13 个，使用这些标签能够完成 JSP 页面的基本功能，减少编码工作。

从功能上可以分为四类：表达式控制标签、流程控制标签、循环标签和 URL 操作标签。

（1）表达式控制标签：out 标签、set 标签、remove 标签、catch 标签。

（2）流程控制标签：if 标签、choose 标签、when 标签、otherwise 标签。

（3）循环标签：forEach 标签、forTokens 标签。

（4）URL 操作标签：import 标签、url 标签、redirect 标签、param 标签。

在 JSP 页面引入核心标签库的代码如下。

```
<%@ taglib prefix="c" uri="http://java.sun.com/jsp/jstl/core" %>
```

专家提醒

使用前首先从官网上下载 JSTL 库，然后在需要使用它的页面引入下载的内容即可，当前最新版本为 1.2.5。

11.3.1 表达式控制标签——out 标签

11.3.1.1 <c:out> 标签的功能

<c:out> 标签主要用来输出数据对象（字符串、表达式）的内容或结果。

11.3.1.2 <c:out> 标签的语法

在使用 Java 脚本输出时常使用的方式代码格式如下。

```
<% out.println(“字符串”)%> 或者 <%= 表达式 %>
```

在 Web 开发中，为了避免暴露逻辑代码会尽量减少页面中的 Java 脚本，可以使用 <c:out> 标签实现以上功能。

```
<c:out value=”字符串”> 或 <c:out value=”EL 表达式”>
```

JSTL 的使用是和 EL 表达式分不开的，EL 表达式虽然可以直接将结果返回给页面，但有时得到的结果为空，<c:out> 有特定的结果处理功能，EL 的单独使用会降低程序的易读性，建议把 EL 的结果放入 <c:out> 标签中。

11.3.1.3 <c:out> 标签的使用案例

```
<%@ page language="java" pageEncoding="UTF-8"%>
<%-- 引入 JSTL 核心标签库 --%>
<%@ taglib prefix="c" uri="http://java.sun.com/jsp/jstl/core"%>
<!DOCTYPE HTML>
<html>
<head>
<title>JSTL: -- 表达式控制标签“out”标签的使用 </title>
</head>
<body>
    <h3>
            <c:out value=" 下面的代码演示了 c:out 的使用，以及在不同属性值状态下的结果。" />
    </h3>
    <hr />
    <ul>
            <%--（1）直接输出了一个字符串。 --%>
```

```
            <li>(1)<c:out value="JSTL 的 out 标签的使用 " /></li>
            <li>(2)<c:out
                    value="<a href='http://www.91isoft.com/'> 点击链接到融创软通网站 </a>" /></li>
            <%--escapeXml="false" 表示 value 值中的 html 标签不进行转义,而是直接输出 --%>
            <li>(3)<c:out
            value="<a href='http://www.91isoft.com/'> 点击链接到融创软通网站 </a>"
            escapeXml="false" /></li>
            <%--(4)字符串中有转义字符,但在默认情况下没有转换。 --%>
            <li>(4)<c:out value="&lt 未使用字符转义 &gt" /></li>
            <%--(5)使用了转义字符 &lt 和 &gt 分别转换成 < 和 > 符号。 --%>
            <li>(5)<c:out value="&lt 使用字符转义 &gt" escapeXml="false"></c:out></li>
            <%--(6)设定了默认值,从 EL 表达式 ${null} 得到空值,所以直接输出设定的默认值。 --%>
            <li>(6)<c:out value="${null}"> 使用了默认值 </c:out></li>
            <%--(7)未设定默认值,输出结果为空。 --%>
            <li>(7)<c:out value="${null}"></c:out></li>
            <%--(8)设定了默认值,从 EL 表达式 ${null} 得到空值,所以直接输出设定的默认值。 --%>
            <li>(8)<c:out value="${null}" default=" 默认值 " /></li>
            <%--(9)未设定默认值,输出结果为空。 --%>
            <li>(9)<c:out value="${null}" /></li>
        </ul>
    </body>
    </html>
```

运行结果如图 11.2 所示。

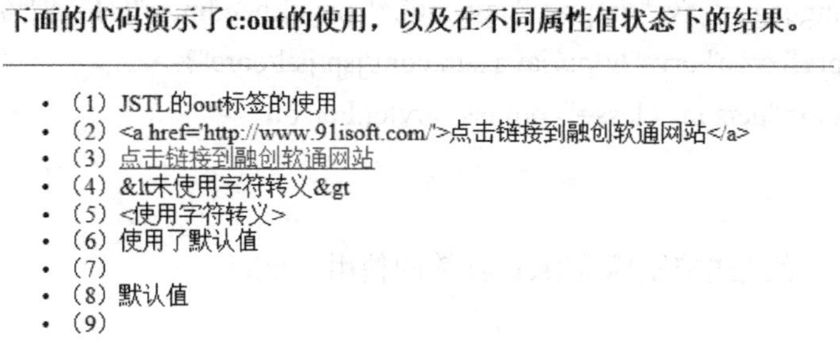

下面的代码演示了c:out的使用，以及在不同属性值状态下的结果。

- （1）JSTL的out标签的使用
- （2）<a href='http://www.91isoft.com/'>点击链接到融创软通网站</a>
- （3）点击链接到融创软通网站
- （4）<未使用字符转义>
- （5）<使用字符转义>
- （6）使用了默认值
- （7）
- （8）默认值
- （9）

图 11.2　运行结果

11.3.2 表达式控制标签——set 标签

11.3.2.1 <c:set> 标签的功能

<c:set> 标签用于把某一个对象存储在指定的域范围内，或者将某一个对象存储到 Map 或者 JavaBean 对象中。

11.3.2.2 <c:set> 标签的语法

<c:set> 标签的编写共有四种语法格式，具体内容如下。

语法 1：存值，把一个值存储在指定的域范围内。

```
<c:set value=" 值 1" var="name1" [scope="page|request|session|application"]/>
```

语法 2：把一个变量名为 name2，值为“值 2”的变量存储在指定的 scope 范围内。

```
<c:set var="name2" [scope="page|request|session|application"]>
    值 2
</c:set>
```

语法 3：把一个值“值 3”赋值给指定的 JavaBean 的属性名，相当于 setter()方法。

```
<c:set value=" 值 3" target="JavaBean 对象 " property=" 属性名 "/>
```

语法 4：把一个值“值 4”赋值给指定的 JavaBean 的属性名。

```
<c:set target="JavaBean 对象 " property=" 属性名 ">
    值 4
</c:set>
```

专家讲解

从功能上分语法 1 和语法 2、语法 3 和语法 4 的效果是一样的，只是 value 值存储的位置不同，至于使用哪个可以根据个人喜好自由选择，语法 1 和语法 2 是向 scope 范围内存储一个值，语法 3 和语法 4 是给指定的 JavaBean 赋值。

11.3.2.3 <c:set> 标签的使用案例

```
<%@ page language="java" import="java.util.*" pageEncoding="UTF-8"%>
<%@ taglib prefix="c" uri="http://java.sun.com/jsp/jstl/core"%>
<jsp:useBean id="person" class="com.iss.servlet.Person" />
<html>
<head>
<title>JSTL: -- 表达式控制标签“set”标签的使用 </title>
</head>
<body>
```

```
	<h3> 代码给出了给指定 scope 范围赋值的示例。</h3>
	<ul>
		<%-- 通过 <c:set> 标签将 data1 的值放入 page 范围中。--%>
		<li> 把 一 个 值 放 入 page 域 中 :<c:set var="data1" value=" 融 创 软 通 "
scope="page" /></li>
		<%-- 使用 EL 表达式从 pageScope 得到 data1 的值。--%>
		<li> 从 page 域中得到值: ${pageScope.data1}</li>
		<%-- 通过 <c:set> 标签将 data2 的值放入 request 范围中。--%>
		<li> 把一个值放入 request 域中 :<c:set var="data2" value=" 融创软通 "
			scope="request" /></li>
		<%-- 使用 EL 表达式从 requestScope 得到 data2 的值。--%>
		<li> 从 request 域中得到值: ${requestScope.data2}</li>
		<%-- 通过 <c:set> 标签将值 name1 的值放入 session 范围中。--%>
		<li> 把一个值放入 session 域中。<c:set value=" 融创软通 " var="name1"
			scope="session"></c:set></li>
		<%-- 使用 EL 表达式从 sessionScope 得到 name1 的值。--%>
		<li> 从 session 域中得到值 :${sessionScope.name1}</li>
		<%-- 把 name2 放入 application 范围中。 --%>
		<li> 把 一 个 值 放 入 application 域 中。<c:set var="name2" scope="applica-
tion"> 融创软通 </c:set></li>
		<%-- 使用 EL 表达式从 application 范围中取值，用 <c:out> 标签输出使得页
面规范化。 --%>
		<li> 使用 out 标签和 EL 表达式嵌套从 application 域中得到值: <c:out
			value="${applicationScope.name2}"> 未 得 到 name 的 值 </
c:out>
		</li>
		<%-- 不指定范围使用 EL 自动查找得到值 --%>
		<li> 未指定 scope 的范围，会从不同的范围内查找得到相应的值: ${data1}、
${data2}、${name1}、${name2}</li>
	</ul>
	<hr />
	<h3> 使用 Java 脚本实现以上功能 </h3>
	<ul>
		<li> 把一个值放入 page 域中。<%
			pageContext.setAttribute("data1", " 融创软通 ");
		%></li>
		<li> 从 page 域中得到值 :<%
```

```
            out.println(pageContext.getAttribute("data1"));
            %></li>
            <li> 把一个值放入 request 域中。<%
                request.setAttribute("data2", " 融创软通 ");
            %></li>
            <li> 从 request 域中得到值 :<%
                out.println(request.getAttribute("data2"));
            %></li>
            <li> 把一个值放入 session 域中。<%
                session.setAttribute("name1", " 融创软通 ");
            %></li>
            <li> 从 session 域中得到值 :<%
                out.println(session.getAttribute("name1"));
            %></li>
            <li><%=session.getAttribute("name1")%></li>
            <li> 把另一个值放入 application 域中。<%
                application.setAttribute("name2", " 融创软通 ");
            %></li>
            <li> 从 application 域中得到值:<%
                out.println(application.getAttribute("name2"));
            %></li>
            <li><%=application.getAttribute("name2")%></li>
            <li> 未指定 scope 的范围,会从不同的范围内查找得到相应的值: <%=page-
Context.findAttribute("data1")%>、
                <%=pageContext.findAttribute("data2")%>、<%=pageContext.findAt-
tribute("name1")%>、
                <%=pageContext.findAttribute("name2")%>
            </li>
        </ul>
        <hr />
        <h3> 操作 JavaBean,设置 JavaBean 的属性值 </h3>
        <c:set target="${person}" property="name"> 融创软通 </c:set>
        <c:set target="${person}" property="age">10</c:set>
        <c:set target="${person}" property="sex"> 男 </c:set>
        <c:set target="${person}" property="home"> 天津 </c:set>
        <ul>
            <li> 使用的目标对象为: ${person}</li>
```

```
                <li> 从 Bean 中获得的 name 值为: <c:out value="${person.name}"></c:out></
li>
                <li> 从 Bean 中获得的 age 值为:<c:out value="${person.age}"></c:out></li>
                <li> 从 Bcan 中获得的 scx 值为:<c:out valuc="${pcrson.sex}"></c:out></li>
                <li> 从 Bean 中获得的 home 值为: <c:out value="${person.home}"></c:out></
li>
        </ul>
        <hr />
        <h3> 操作 Map</h3>
        <%
                Map map = new HashMap();
                request.setAttribute("map", map);
        %>
        <%-- 将 data 对象的值存储到 map 集合中 --%>
        <c:set property="data" value=" 融创软通 " target="${map}" />
        ${map.data}
    </body>
    </html>
```

JSP 页面中使用到的 Person 类的代码如下。

```
public class Person {
    private String age;
    private String home;
    private String name;
    private String sex;
// 省略 setter 和 getter
}
```

运行结果如图 11.3 所示。

代码给出了给指定scope范围赋值的示例。

- 把一个值放入page域中:
- 从page域中得到值：融创软通
- 把一个值放入request域中:
- 从request域中得到值：融创软通
- 把一个值放入session域中。
- 从session域中得到值:融创软通
- 把一个值放入application域中。
- 使用out标签和EL表达式嵌套从application域中得到值： 融创软通
- 未指定scope的范围，会从不同的范围内查找得到相应的值：融创软通、融创软通、融创软通、融创软通

使用Java脚本实现以上功能

- 把一个值放入page域中。
- 从page域中得到值:融创软通
- 把一个值放入request域中。
- 从request域中得到值:融创软通
- 把一个值放入session域中。
- 从session域中得到值:融创软通
- 融创软通
- 把另一个值放入application域中。
- 从application域中得到值：融创软通
- 融创软通
- 未指定scope的范围，会从不同的范围内查找得到相应的值： 融创软通、融创软通、融创软通、融创软通

操作JavaBean，设置JavaBean的属性值

- 使用的目标对象为：com.iss.servlet.Person@5c1dbe2d
- 从Bean中获得的name值为：融创软通
- 从Bean中获得的age值为：10
- 从Bean中获得的sex值为：男
- 从Bean中获得的home值为：天津

操作Map

融创软通

图 11.3　运行结果

11.3.3　表达式控制标签——remove 标签

11.3.3.1　<c:remove> 标签的功能

<c:remove> 标签主要用来从指定的 JSP 范围内移除指定的变量。

11.3.3.2　<c:remove> 标签的语法

```
<c:remove var=" 变量名 " [scope="page|request|session|application"]/>
```

其中 var 属性是必须的，scope 可以省略。

11.3.3.3　<c:remove> 标签的使用案例

```
<ul>
    <c:set var="name" scope="session"> 融创软通 </c:set>
    <c:set var="age" scope="session">25</c:set>
    <li><c:out value="${sessionScope.name}"></c:out></li>
    <li><c:out value="${sessionScope.age}"></c:out></li>
```

```
    <%-- 使用 remove 标签移除 age 变量 --%>
    <c:remove var="age" />
    <li><c:out value="${sessionScope.name}"></c:out></li>
    <li><c:out valuc="${scssionScopc.agc}"></c:out></li>
</ul>
```

11.3.4 表达式控制标签——catch 标签

11.3.4.1 <c:catch> 标签的功能

<c:catch> 标签用于捕获嵌套在标签体中的内容抛出的异常。

11.3.4.2 <c:catch> 标签的语法

```
<c:catch [var="varName"]> 容易产生异常的代码 </c:catch>
```

其中 var 属性用于标识 <c:catch> 标签捕获的异常对象，它将保存在 page Web 域中。

11.3.4.3 <c:catch> 标签的使用案例

```
<%-- 把容易产生异常的代码放在 <c:catch></c:catch> 中，
    自定义一个变量 errorInfo 用于存储异常信息 --%>
    <c:catch var="errorInfo">
        <%-- 实现了一段异常代码，向一个不存在的 JavaBean 中插入一个值 --%>
        <c:set target="person" property="hao"></c:set>
    </c:catch>
    <%-- 用 EL 表达式得到 errorInfo 的值，并使用 <c:out> 标签输出 --%>
    异常：
    <c:out value="${errorInfo}" />
    <br /> 异常 errorInfo.getMessage：
    <c:out value="${errorInfo.message}" />
    <br /> 异常 errorInfo.getCause：
    <c:out value="${errorInfo.cause}" />
    <br /> 异常 errorInfo.getStackTrace：
    <c:out value="${errorInfo.stackTrace}" />
```

运行结果如图 11.4 所示。

catch标签实例

异常： javax.servlet.jsp.JspTagException: Invalid property in <set>: "hao"
异常 errorInfo.getMessage： Invalid property in <set>: "hao"
异常 errorInfo.getCause：
异常 errorInfo.getStackTrace： [Ljava.lang.StackTraceElement;@75058b5

图 11.4 运行结果

11.3.5 流程控制标签——if 标签

11.3.5.1 <c:if> 标签的功能

<c:if> 标签和程序中的 if 语句作用相同，用来实现条件控制。

11.3.5.2 <c:if> 标签的语法

```
<c:if test="testCondition" [var="varName"] [scope="{page|request|session|application}"]>
        标签体内容
    </c:if>
```

参数说明如下。

（1）test 属性用于存放判断的条件，一般使用 EL 表达式来编写。

（2）var 属性用来存放判断的结果，类型为 true 或 false。

（3）scope 属性用来指定 var 属性存放的范围。

11.3.5.3 <c:if> 标签的使用案例

```
<h4>if 标签示例 </h4>
    <hr>
<form action="index.jsp" method="post">
        <input type="text" name="uname" value="${param.uname}">
    <input type="submit" value=" 登录 ">
    </form>
    <c:if test="${param.uname=='admin'}" var="adminchock">
        <c:out value=" 管理员欢迎您！ " />
    </c:if>
    ${adminchock}
```

运行结果如图 11.5 所示。

if标签示例

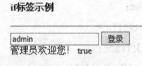

admin 登录
管理员欢迎您！ true

图 11.5 运行结果

11.3.6 流程控制标签——choose 标签、when 标签、otherwise 标签

11.3.6.1 <c:choose>、<c:when> 和 <c:otherwise> 标签的功能

<c:choose>、<c:when> 和 <c:otherwise> 这三个标签在通常情况下是一起使用的，<c:choose> 标签作为 <c:when> 和 <c:otherwise> 标签的父标签来使用。

11.3.6.2 语法

```
<c:choose>
   <c:when test=" 条件 1">
           // 业务逻辑 1
   <c:when>
     <c:when test=" 条件 2">
           // 业务逻辑 2
   <c:when>
     <c:when test=" 条件 n">
           // 业务逻辑 n
   <c:when>
   <c:otherwise>
           // 业务逻辑
     </c:otherwise>
</c:choose>
```

11.3.6.3 使用案例

```
<h4>choose 及其嵌套标签示例 </h4>
  <hr />
  <c:set var="score" value="85" />
  <c:choose>
         <c:when test="${score>=90}">
     您的成绩为优秀！
   </c:when>
         <c:when test="${score>70 && score<90}">
     您的成绩为良好！
   </c:when>
         <c:when test="${score>60 && score<70}">
     您的成绩为及格
   </c:when>
         <c:otherwise>
     对不起，您没有通过考试！
   </c:otherwise>
  </c:choose>
```

11.3.7 循环标签——forEach 标签

11.3.7.1 <c:forEach> 标签的功能

<c:forEach> 标签根据循环条件遍历集合（Collection）中的元素。

11.3.7.2 <c:forEach> 标签的语法

```
<c:forEach
      var="name"
      items="Collection"
      varStatus="StatusName"
      begin="begin"
      end="end"
      step="step">
  本体内容
</c:forEach>
```

参数说明如下。

（1）var 设定变量名用于存储从集合中取出的元素。

（2）items 指定要遍历的集合。

（3）varStatus 设定变量名，该变量用于存放集合中元素的信息。

（4）begin、end 用于指定遍历的起始位置和终止位置（可选）。

（5）step 指定循环的步长。

11.3.7.3 <c:forEach> 标签属性

循环标签属性说明见表 11.1。

表 11.1 循环标签属性说明

属性名称	是否支持 EL 表达式	属性类型	是否必须	默认值
var	NO	String	是	无
items	YES	Arrays Collection Iterator Enumeration Map String []args	是	无
begin	YES	int	否	0
end	YES	int	否	集合中最后一个元素
step	YES	int	否	1
varStatus	NO	String	否	无

其中 varStatus 有四个状态属性，见表 11.2。

表 11.2 varStatus 的四个状态

属性名	类型	说明
index	int	当前循环的索引值
count	int	循环的次数
first	boolean	是否为第一个位置
last	boolean	是否为最后一个位置

11.3.7.4 <c:forEach> 使用案例

```
<h4>
        <c:out value="forEach 实例 " />

</h4>
  <%
        List<String> list = new ArrayList<String>();
        list.add(0, " 百度 ");
        list.add(1, " 阿里 ");
        list.add(2, " 京东 ");
        list.add(3, " 腾讯 ");
        list.add(4, " 华为 ");
        request.setAttribute("list", list);
  %>
  <B><c:out value=" 不指定 begin 和 end 的迭代：" /></B>
  <br>
  <c:forEach var="it" items="${list}">
    <c:out value="${it}" />
        <br />
  </c:forEach>
  <B><c:out value=" 指定 begin 和 end 的迭代：" /></B>
  <br>
  <c:forEach var="it" items="${list}" begin="1" end="3" step="2">
    <c:out value="${it}" />
        <br />
  </c:forEach>
  <B><c:out value=" 输出整个迭代的信息：" /></B>
```

```
<br>
<c:forEach var="it" items="${list}" begin="3" end="4" varStatus="s"
        step="1">
    <c:out value="${it}" /> 的四种属性:<br>
    所在位置,即索引:<c:out value="${s.index}" />
        <br>
    总共已迭代的次数:<c:out value="${s.count}" />
        <br>
    是否为第一个位置:<c:out value="${s.first}" />
        <br>
    是否为最后一个位置:<c:out value="${s.last}" />
        <br>
</c:forEach>
```

运行结果如图 11.6 所示。

forEach实例

不指定begin和end的迭代：
百度
阿里
京东
腾讯
华为
指定begin和end的迭代：
阿里
腾讯
输出整个迭代的信息：
腾讯的四种属性：
所在位置，即索引：3
总共已迭代的次数：1
是否为第一个位置：true
是否为最后一个位置：false
华为的四种属性：
所在位置，即索引：4
总共已迭代的次数：2
是否为第一个位置：false
是否为最后一个位置：true

图 11.6　运行结果

11.3.8　循环标签——forTokens 标签

11.3.8.1　<c:forTokens> 标签的功能

<c:forTokens> 标签用于浏览字符串，并根据指定的字符将字符串截取。

11.3.8.2　<c:forTokens> 标签的语法

```
<c:forTokens items="strigOfTokens"
        delims="delimiters"
        [var="name"
```

```
        begin="begin"
        end="end"
        step="len"
        varStatus="statusName"] >
    主体内容
</c:forTokens>
```

参数说明如下。

（1）items 指定被迭代的字符串。

（2）delims 指定使用的分隔符。

（3）var 指定用来存储遍历对象的临时对象。

（4）begin 指定遍历开始的位置（int 型从取值 0 开始）。

（5）end 指定遍历结束的位置（int 型，默认集合中最后一个元素）。

（6）step 遍历的步长（大于 0 的整型）。

（7）varStatus 存储遍历到的成员的状态信息。

11.3.8.3 <c:forTokens> 使用案例

```
<h4>
        <c:out value="forToken 实例 " />
</h4>
<hr />
<c:forTokens var="str" items=" 融、创、软、通、欢、迎、你 " delims="、">
        <c:out value="${str}"></c:out>  
</c:forTokens>
```

运行结果如图 11.7 所示。

forToken实例

融 创 软 通 欢 迎 你

图 11.7　运行结果

11.3.9 URL 操作标签——import 标签

11.3.9.1 <c:import> 标签的功能

<c:import> 标签可以把其他静态或动态文件包含到本 JSP 页面，与 <jsp:include> 的区别为 <jsp:include> 只能包含同一个 Web 应用中的文件，而 <c:import> 可以包含其他 Web 应用中的文件，甚至是网络上的资源。

11.3.9.2 <c:import> 标签的语法

```
<c:import
   url="url"
   varReader="name"
   [context="context"]
   [charEncoding="encoding"]/>
```

参数说明如下。

(1)URL 为资源的路径，当引用的资源不存在时系统会抛出异常，因此该语句应该放在 <c:catch></c:catch> 语句块中捕获。

(2)引用资源有绝对路径和相对路径两种方式。

使用绝对路径的示例：<c:import url="http://www.baidu.com">。

使用相对路径的示例：<c:import url="aa.txt">，aa.txt 需要与当前页面存放在同一文件目录。

(3)如果以“/”开头则表示程序部署的根目录。例如：Tomcat 应用程序的根目录文件夹为 webapps，因此从程序根目录导入文件 bb.txt 的编写方式为 <c:import url="/bb.txt">。

如果访问 webapps 管理文件夹中的其他 Web 应用就要用 context 属性。

(4)context 属性用于在访问其他 Web 应用的文件时，指定根目录。例如，访问 root 下的 index.jsp 的实现代码为 <c:import url="/index.jsp" context="/root">，等同于 webapps/root/index.jsp。

(5)var、scope、charEncoding、varReader 是可选属性。

11.3.9.3 <c:import> 标签的使用案例

```
<h4>
        <c:out value="import 实例 " />
   </h4>
   <hr />
   <h4>
        <c:out value=" 绝对路径引用的实例 " />
   </h4>
   <c:catch var="error1">
        <c:import url="http://www.baidu.com" charEncoding="utf-8" />
   </c:catch>
   ${error1}
```

11.3.10 URL 操作标签——url 标签

11.3.10.1 <c:url> 标签的功能

<c:url> 标签用于在 JSP 页面中构造一个 URL 地址，其主要目的是实现 URL 重写。

11.3.10.2 <c:url> 标签的语法

```
<c:url
   value="value"
   [var="name"]
   [scope="page|request|session|application"]
   [context="context"]>
       <c:param name=" 参数名 " value=" 值 ">
</c:url>
```

11.3.10.3 <c:url> 标签使用案例

```
<c:out value="url 标签使用 "></c:out>
<h4> 使用 url 标签生成一个动态的 url，并把值存入 session 中 .</h4>
<hr />
<c:url value="http://www.baidu.com" var="url" scope="session">
</c:url>
<a href="${url}"> 百度首页（不带参数）</a>
<hr />
<h4> 配合 &lt;c:param&gt; 标签给 url 加上指定参数及参数值，生成一个动态的 url
然后存储到 paramUrl 变量中
</h4>
<c:url value="http://www.baidu.com" var="paramUrl">
        <c:param name="userName" value=" 融创软通 " />
        <c:param name="pwd">123456</c:param>
</c:url>
<a href="${paramUrl}"> 百度首页（带参数）</a>
```

运行结果如图 11.8 所示。

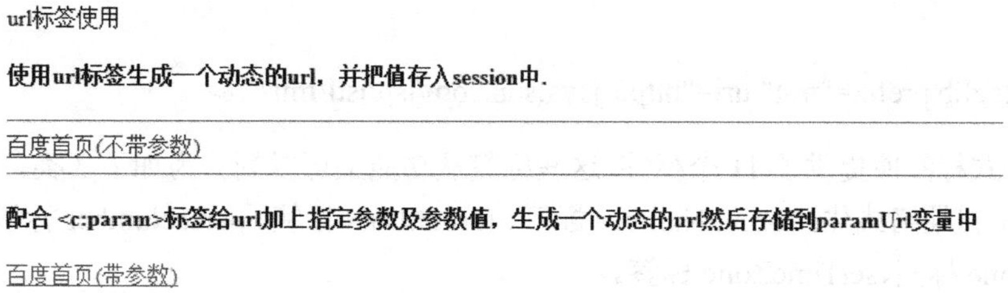

url标签使用

使用url标签生成一个动态的url，并把值存入session中.

百度首页(不带参数)

配合 <c:param>标签给url加上指定参数及参数值，生成一个动态的url然后存储到paramUrl变量中

百度首页(带参数)

图 11.8 运行结果

11.3.11 URL 操作标签——redirect 标签

11.3.11.1 <c:redirect> 标签的功能

<c:redirect> 标签用来实现请求的重定向，同时可以配合使用 <c:param> 标签在 url 中加入指定的参数。

11.3.11.2 <c:redirect> 标签的语法

```
<c:redirect url="url"[context="context"]>
<c:param name="name1" value="value1">
</c:redirect>
```

参数说明如下。

（1）url 指定重定向页面的地址，可以是一个 string 类型的绝对地址或相对地址。

（2）context 用于导入其他 Web 应用中的页面。

11.3.11.3 <c:redirect> 标签使用案例

```
<c:redirect url="http://www.baidu.com">
        <%-- 在重定向时使用 <c:param> 标签为 URL 添加了两个参数 --%>
        <c:param name="uname"> 融创软通 </c:param>
        <c:param name="password">12345678</c:param>
</c:redirect>
```

关于 JSTL 核心标签库掌握以上内容基本上就可以应付开发了。

11.4 格式化标签库

JSTL 标签提供了对国际化（I18N）的支持，可以根据发出请求的客户端地域的不同来显示不同的语言，同时还提供了格式化数据和日期的方法。

实现这些功能需要 I18N 格式标签库（I18N-capable formation tags liberary）；引入该标签库的方法为

```
<%@ taglib prefix="fmt" uri="http://java.sun.com/jsp/jstl/fmt" %>
```

I18N 格式标签库提供了 11 个标签，这些标签从功能上可以划分为如下 3 类。

（1）数字日期格式化：formatNumber 标签、formatData 标签、parseNumber 标签、parseDate 标签、timeZone 标签、setTimeZone 标签。

（2）读取消息资源：bundle 标签、message 标签、setBundle 标签。

（3）国际化：setlocale 标签、requestEncoding 标签。

11.4.1 格式化日期标签——<fmt:formatDate /> 标签

11.4.1.1 <fmt:formatDate /> 标签的功能

<fmt:formatDate /> 标签用来将日期类型转换为以字符串形式表现的日期。

11.4.1.2 <fmt:formatDate /> 标签的语法

```
<fmt:formatDate value="number" [type={time|date|both}]
[pattern="pattern"]
[dateStyle="{default|short|medium|long|full}"]
[timeStyle="{default|short|medium|long|full}"]
[timeZone="timeZone"]
[var="varname"]
[scope="page|request|session|application"]
/>
```

11.4.1.3 <fmt:formatDate /> 标签使用实例

```
<%
        pageContext.setAttribute("today", new Date());
  %>
  <fmt:formatDate value="${today}" />
  <fmt:formatDate value="${today}" type="time" />
  <fmt:formatDate value="${today}" type="both" />
  <fmt:formatDate value="${today}" dateStyle="short" />
  <fmt:formatDate value="${today}" dateStyle="medium" />
  <fmt:formatDate value="${today}" dateStyle="long" />
  <fmt:formatDate value="${today}" dateStyle="full" />
  <fmt:formatDate value="${today}" pattern="yyyy/MM/dd  HH:mm:ss" />
  <fmt:formatDate value="${today}" pattern="yyyy/MM/dd  HH:mm:ss" var="d" />
  ${d}
```

运行结果如图 11.9 所示。

```
2017-10-15
22:50:46
2017-10-15 22:50:46
17-10-15
2017-10-15
2017年10月15日
2017年10月15日 星期日
2017/10/15 22:50:46
2017/10/15 22:50:46
```

图 11.9 运行结果

11.4.2 格式化数字标签——<fmt:formatNumber> 标签

11.4.2.1 <fmt:formatNumber> 标签的功能

<fmt:formatNumber> 标签根据区域或定制的方式将数字格式化成数字、货币或百分比。

11.4.2.2 <fmt:formatNumber> 标签的语法

```
<fmt:formatNumber value="number" [type={number|currency|percent|}]
 [pattern="pattern"]
 [currencyCode="currencyCode"]
 [currentSymbol="currentSymbol"]
 [groupingUsec="{true|false}"]
 [maxIntergerDigits="maxIntergerDigits"]
 [minIntergerDigits="minIntergerDigits"]
 [maxFractionDigits="maxFractionDigits"]
 [minFractionDigits="minFractionDigits"]
 [var="varname"]
 [scope="page|request|session|application"]
 />
```

11.4.2.3 <fmt:formatNumber> 标签使用实例

```
<%@page language="java" contentType="text/html;charset=utf-8"%>
<%@ taglib uri="http://java.sun.com/jsp/jstl/fmt" prefix="fmt" %>
<!DOCTYPE html>
<html>
 <head>
   <title>FormatNumber 标签使用 </title>
 </head>
 <body>
   <h1>FormatNumber 标签使用 </h1>
   <fmt:setLocale value="fr_fr" />
   France:<fmt:formatNumber value="123456789.012"/>
   <fmt:setLocale value="zh_cn" />
   China:<fmt:formatNumber value="123456789.012"/>
   <fmt:setLocale value="de_de" />
   Germany:<fmt:formatNumber value="123456789.012"/>
 </body>
</html>
```

11.4.3 数字转换标签——<fmt:parseNumber /> 标签

11.4.3.1 <fmt:parseNumber /> 标签的功能

<fmt:parseNumber /> 标签用来将字符串类型的数字、货币或百分比转换成数字类型。

11.4.3.2 <fmt:parseNumber /> 标签的语法

```
<fmt:parseNumber value="numberString" [type={number|currency|percent|}]
 [pattern="pattern"]
 [parseLocale="parseLocale"]
 [integerOnly="{false|true}"]
 [var="varname"]
 [scope="page|request|session|application"]
 />
```

11.4.4 字符串转日期标签——<fmt:parseDate /> 标签

11.4.4.1 <fmt:parseDate /> 标签的功能

<fmt:parseDate /> 标签用来将字符串类型的时间或日期转换成日期时间类型。

11.4.4.2 <fmt:parseDate /> 标签的语法

```
<fmt:parseDate value="date" [type={time|date|both}]
 [pattern="pattern"]
 [dateStyle="{default|short|medium|long|full}"]
 [timeStyle="{default|short|medium|long|full}"]
 [timeZone="timeZone"]
 [var="varname"]
 [scope="page|request|session|application"]
 />
```

11.4.5 设置时区标签——<fmt:setTimeZone /> 标签

11.4.5.1 <fmt:setTimeZone /> 标签的功能

<fmt:setTimeZone /> 标签用来设置默认时区或将时区存储到属性范围中。

11.4.5.2 <fmt:setTimeZone /> 标签的语法

```
<fmt:setTimeZone value="timezone" [var="varname"] [scope="{page|request|session|application}"] />
```

11.4.6 <fmt:timeZone /> 标签

11.4.6.1 <fmt:timeZone /> 标签的功能

<fmt:timeZone /> 标签用来暂时设定时区。

11.4.6.2 <fmt:timeZone /> 标签的语法

```
<fmt:timeZone value="timeZone">
主体内容
</fmt:timeZone>
```

11.4.7 <fmt:setLocale /> 标签

11.4.7.1 <fmt:setLocale /> 标签的功能

<fmt:setLocale /> 标签用来设定用户的区域语言。

11.4.7.2 <fmt:setLocale /> 标签的语法

```
<fmt:setLocale value="locale" [variant="variant"] [scope="{page|request|session|applica-
tion}"] />
```

11.4.8 <fmt:requestEncoding /> 标签

11.4.8.1 <fmt:requestEncoding /> 标签的功能

<fmt:requestEncoding /> 标签设定接收的字符串的编码格式。

11.4.8.2 <fmt:requestEncoding /> 标签的语法

```
<fmt:requestEncoding value="charsetName" />
```

11.4.9 <fmt:setBundle /> 标签

11.4.9.1 <fmt:setBundle /> 标签的功能

<fmt:setBundle /> 标签用来设定默认的数据来源,也可以将其存储到一定范围中供需要时使用。

11.4.9.2 <fmt:setBundle /> 标签的语法

```
<fmt:setBundle basename="basename" [var="varname"] [scope="{page|request|session|ap-
plication}"] />
```

11.4.10 <fmt:message /> 标签

11.4.10.1 <fmt:message /> 标签的功能

<fmt:message /> 标签用来从指定的资源文件中通过索引取得值。

11.4.10.2 <fmt:message /> 标签的语法

```
<fmt:message key="messageKey" [bundle="resourceBundle"] [var="varname"] [scope="{page|request|session|application}"] />
```

11.4.11 <fmt:param /> 标签

11.4.11.1 <fmt:param /> 标签的功能

<fmt:param /> 标签用来传递参数（在从资源文件中取得信息时，可能需要动态设定参数的情况下）。

11.4.11.2 <fmt:param /> 标签的语法

```
<fmt:param value="messageParameter" > 有本体内容
参数
</fmt:param>
```

11.4.12 <fmt:bundle /> 标签

11.4.11.1 <fmt:bundle /> 标签的功能

<fmt:bundle /> 标签用来设定数据来源。

11.4.11.2 <fmt:bundle /> 标签的语法

```
<fmt:bundle basename="basename" [prefix="prefix"] >
本体内容 <fmt:message>
</fmt:bundle>
```

小结

1. JSTL 标签用于访问数据库、操作、设置和删除作用域变量和国际化标签。
2. 核心标签库是最常用的标签库。
3. 通用标签用于设置、删除和显示表达式的输出结果。
4. 条件标签用于有条件地执行标签。
5. if 和 choose 标签用于有条件地执行标签。
6. 迭代标签用于多次执行标签体。
7. forEach 和 forTokens 是迭代标签。
8. I18N 与格式化标签库可用于创建国际化和标准化的 Web 应用程序。
9. SQL 标签用于访问和更新数据库。
10. SQL 标签适用于基于 Web 的小型应用程序。

经典面试题

1. 什么是 JSTL？
2. 使用 JSTL 需要引用哪些 jar 包？
3. JSP 如何将 EL 表达式读取出的 date 型数据转换成 string 型？
4. JSP 中怎样引用 JSTL 标签？
5. JSTL 标准标签库包括什么标签？
6. 如何在 JSP 中用 JSTL 实现一个分页？
7. 一个实体类集合如何用 JSTL 遍历出来？
8. JSP 中的 JSTL 与 EL 表达式有什么区别？
9. JSTL 标签怎么接收从其他 JSP 传递过来的参数？
10. 举例说明 JSTL 的函数 functions 的用法？

跟我上机

1. 编写一个程序，使用 foreach 标签，在 JSP 页面中显示产品的名称及成本。（提示产品的名称和成本可以存储在数组变量中）

2. 编写一个程序，使用格式化标签，分别显示瑞士、美国和中国样式的日期和数字。（提示要求使用 formatdata 和 formatnumber 标签）

3. 创建一个 JSP 页面，允许用户选择显示数据所需的语言，根据所选语言接受并显示用户输入的数字和日期。

4. 创建一个 HTML 页面，接收商品详细信息，如商品名称、类型、品牌、价格和说明。接收商品详细信息后，HTML 页需要调用一个 JSP 页面，将这些详细信息插入数据库，并在表上显示各项记录。

5. 使用 JSTL+MySQ1 数据库制作如下图功能的投票系统。

168 英雄人物投票

○	1	刘备	8	4.76%
○	2	关羽	11	6.55%
○	3	张飞	102	60.71%
◉	4	赵云	47	27.98%

投票

第 12 章　Filter 和 Listener

本章要点(学会后请在方框里打钩):

- □ 了解什么是过滤器(Filter)
- □ 了解 Filter 的工作原理
- □ 掌握 Filter 在项目中的应用
- □ 掌握什么是监听器(Listener)
- □ 掌握监听器在项目中的应用

12.1 Filter 简介

Filter 也称过滤器，是 Servlet 技术中最激动人心的技术，Web 开发人员通过 Filter 技术，对 Web 服务器管理的所有 Web 资源，如 JSP、Servlet、静态图片文件或静态 HTML 文件等进行拦截，从而实现一些特殊功能。例如实现 URL 级别的权限访问控制、过滤敏感词汇、压缩响应信息等高级功能。

过滤器是一个服务器端的组件，它可以截取用户端的请求与响应信息，并对这些信息进行过滤。

Servlet API 中提供了一个 Filter 接口，在开发 Web 应用时，如果编写的 Java 类实现了这个接口，则把这个 Java 类称为过滤器 Filter。通过 Filter 技术，开发人员可以实现用户在访问某个目标资源之前，对访问的请求和响应进行拦截。

Servlet 过滤器是在 Java Servlet 规范 2.3 中定义的，它能够对 Servlet 容器的请求和响应对象进行检查和修改。

Servlet 过滤器本身并不产生请求和响应对象，它只能提供过滤作用。Servlet 过滤器能够在 Servlet 被调用之前检查 Request 对象，修改 Request Header 以及 Request 中的内容；也可以在 Servlet 被调用之后检查 Response 对象，修改 Response Header 以及 Response 中的内容。

12.1.1 Filter 工作原理

图 12.1 为过滤器（Filter）的工作原理。Filter 接口中有一个 doFilter 方法，编写好 Filter 后，Web 服务器在调用 doFilter 方法时，会传递一个 FilterChain 对象进来，FilterChain 对象是 Filter 接口中最重要的一个对象，它也提供了一个 doFilter 方法，开发人员可以根据需求决定，如果调用该方法，则 Web 服务器就会调用 Web 资源的 Service 方法，即 Web 资源就会被访问，否则 Web 资源不会被访问。

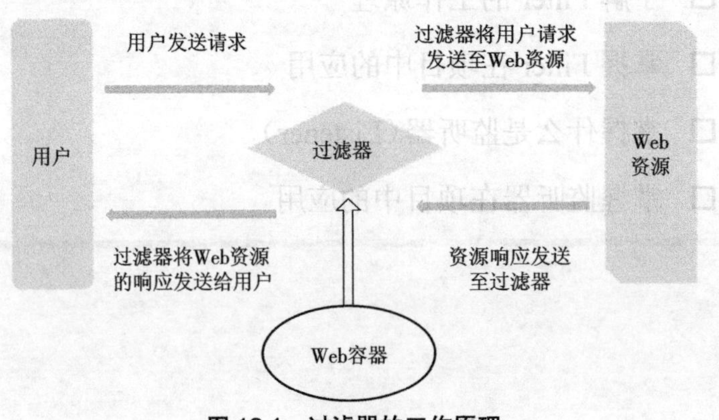

图 12.1 过滤器的工作原理

12.1.2 Filter 开发入门

Filter 开发有以下两个步骤。

（1）编写 Java 类实现 Filter 接口，并实现 doFilter 方法。

（2）在 web.xml 文件中使用 <filter> 和 <filter-mapping> 元素对编写的 Filter 类进行注册，并设置它所能拦截的资源。

实例 12.1：过滤器

```
@WebFilter("/*")
public class FilterDemo1 implements Filter {
    @Override
    public void init(FilterConfig filterConfig) throws ServletException {
            System.out.println("---- 过滤器初始化 ----");
    }
    @Override
    public void doFilter(ServletRequest request, ServletResponse response, FilterChain
chain)
                    throws IOException, ServletException {
            // 对 request 和 response 进行一些预处理
            request.setCharacterEncoding("UTF-8");
            response.setCharacterEncoding("UTF-8");
            response.setContentType("text/html;charset=UTF-8");
            System.out.println("FilterDemo01 执行前!!! ");
            chain.doFilter(request, response); // 让目标资源执行，放行
            System.out.println("FilterDemo01 执行后!!! ");
    }
    @Override
    public void destroy() {
            System.out.println("---- 过滤器销毁 ----");
    }
}
```

运行结果如图 12.2 所示。

```
信息: Creation of SecureRandom in
----过滤器初始化----
十二月13, 2017 11:32:26 上午org.ap
信息: Starting ProtocolHandler ["
十二月13, 2017 11:32:26 上午org.ap
信息: Starting ProtocolHandler ["
十二月13, 2017 11:32:26 上午org.ap
信息: Server startup in 2366 ms
FilterDemo01执行前!!!
FilterDemo01执行后!!!
```

图 12.2　运行结果

12.1.3　Filter 链

在一个 Web 应用中,可以开发编写多个 Filter,这些 Filter 组合起来称为一个 Filter 链,如图 12.3 所示。

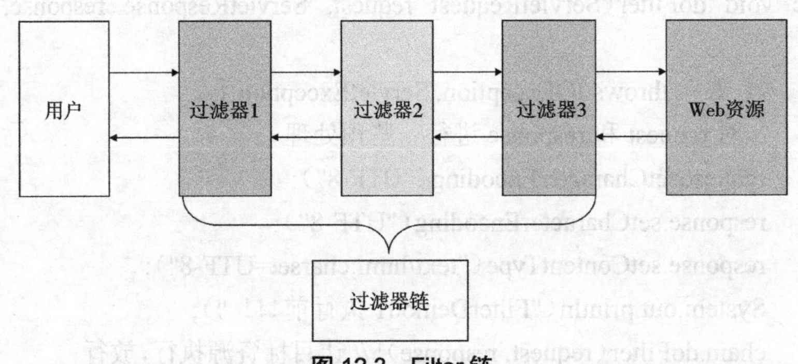

图 12.3　Filter 链

Web 服务器根据 Filter 在 web.xml 文件中的注册顺序,决定先调用哪个 Filter,当第一个 Filter 的 doFilter 方法被调用时, Web 服务器会创建一个代表 Filter 链的 FilterChain 对象传递给该方法。在 doFilter 方法中,开发人员如果调用了 FilterChain 对象的 doFilter 方法,则 Web 服务器会检查 FilterChain 对象中是否还有 Filter:如果有,则调用第二个 filter;如果没有,则调用目标资源。

12.1.4　Filter 的生命周期

1.Filter 的创建

Filter 的创建和销毁由 Web 服务器负责。Web 应用程序启动时, Web 服务器将创建 Filter 的实例对象,并调用 init 方法,完成对象的初始化功能,从而为后续的用户请求做好拦截的准备工作, Filter 对象只会创建一次, init 方法也只会执行一次。通过 init 方法的参数,可获得代表当前 Filter 配置信息的 FilterConfig 对象。

2.Filter 的销毁

Web 容器调用 destroy 方法销毁 Filter。destroy 方法在 Filter 的生命周期中仅执行一次。

在 destroy 方法中，可以释放过滤器使用的资源。

3.FilterConfig 接口

用户在配置 Filter 时，可以使用 <init-param> 为 Filter 配置一些初始化参数，当 Web 容器实例化 Filter 对象，调用其 init 方法时，会把封装了 Filter 初始化参数的 FilterConfig 对象传递进来。因此开发人员在编写 Filter 时，通过 FilterConfig 对象的方法，可获得如下内容。

（1）String getFilterName（）：得到 Filter 的名称。

（2）String getInitParameter（String name）：返回在部署描述中指定名称的初始化参数的值，如果不存在则返回 null。

（3）Enumeration getInitParameterNames（）：返回过滤器的所有初始化参数名字的枚举集合。

（4）public ServletContext getServletContext（）：返回 Servlet 上下文对象的引用。

实例 12.2：利用 FilterConfig 得到 Filter 配置信息

```
@WebFilter(value = "/*", initParams = { @WebInitParam(name = "name", value = " 融创软通 "),
            @WebInitParam(name = "like", value = "Java 开发 ") })
public class FilterDemo2 implements Filter {
    @Override
    public void init(FilterConfig filterConfig) throws ServletException {
            System.out.println("---- 过滤器初始化 ----");
            // 得到过滤器的名字
            String filterName = filterConfig.getFilterName();
            // 得到在 web.xml 文件中配置的初始化参数
            String initParam1 = filterConfig.getInitParameter("name");
            String initParam2 = filterConfig.getInitParameter("like");
            // 返回过滤器的所有初始化参数名字的枚举集合。
            Enumeration<String> initParameterNames = filterConfig.getInitParameterNames();
            System.out.println(filterName);
            System.out.println(initParam1);
            System.out.println(initParam2);
            while (initParameterNames.hasMoreElements()) {
                    String paramName = (String) initParameterNames.nextElement();
                    System.out.println(paramName);
            }
    }
    @Override
```

```
    public void doFilter(ServletRequest request, ServletResponse response, FilterChain chain)
                throws IOException, ServletException {
            System.out.println("FilterDemo02 执行前!!! ");
            chain.doFilter(request, response); // 让目标资源执行,放行
            System.out.println("FilterDemo02 执行后!!! ");
    }
    @Override
    public void destroy() {
            System.out.println("---- 过滤器销毁 ----");
    }
}
```

运行结果如图 12.4 所示。

```
-----过滤器初始化-----
com.isoft.filter.FilterDemo2
融创软通
Java开发
like
name
十二月 13, 2017 11:39:58 上午 org.a
信息: Reloading Context with nam
FilterDemo02执行前!!!
FilterDemo02执行后!!!
```

图 12.4　运行结果

12.1.5 Filter 的部署

Filter 的部署分为如下两个步骤。

1. 注册 Filter

开发好 Filter 之后,需要在 web.xml 文件中进行注册,这样才能够被 Web 服务器调用。

实例 12.3:在 web.xml 文件中注册 Filter

```
<filter>
        <description>FilterDemo02 过滤器 </description>
        <filter-name>FilterDemo02</filter-name>
        <filter-class>com.isoft.web.filter.FilterDemo02</filter-class>
        <!-- 配置 FilterDemo02 过滤器的初始化参数 -->
<init-param>
        <description> 配置 FilterDemo02 过滤器的初始化参数 </description>
        <param-name>name</param-name>
```

```
            <param-value> 融创软通 IT 学院 </param-value>
        </init-param>
        <init-param>
            <dcscription> 配置 FiltcrDcmo02 过滤器的初始化参数 </description>
            <param-name>like</param-name>
            <param-value>Java 开发 </param-value>
        </init-param>
    </filter>
```

专家讲解

<description>：用于添加描述信息，该元素的内容可为空，<description> 可以不配置。

<filter-name>：用于为过滤器指定一个名字，该元素的内容不能为空。

<filter-class>：元素用于指定过滤器的完整的限定类名。

<init-param>：元素用于为过滤器指定初始化参数，它的子元素 <param-name> 指定参数的名字，<param-value> 指定参数的值。在过滤器中，可以使用 FilterConfig 接口对象来访问初始化参数。如果过滤器不需要指定初始化参数，那么 <init-param> 元素可以不配置。

2. 映射 Filter

在 web.xml 文件中注册了 Filter 之后，还要在 web.xml 文件中映射 Filter。

```
<!-- 映射过滤器 -->
<filter>
    <filter-name>FilterDemo02</filtername>
    <filter-class>com.iss.iedu.filter.FilterDemo02</filter-class>
</filter>
<filter-mapping>
    <filter-name>FilterDemo02</filtername>
    <url-pattern>/*</url-pattern>
    <dispatcher>REQUEST</dispatcher>
    <dispatcher>FORWARD</dispatcher>
    <dispatcher>INCLUDE</dispatcher>
    <dispatcher>EXCEPTION</dispatcher>
</filter-mapping>
```

<filter-mapping> 元素用于设置一个 Filter 所负责拦截的资源。一个 Filter 拦截的资源可通过两种方式来指定：Servlet 名称和资源访问的请求路径。

<filter-name> 子元素用于设置 Filter 的注册名称，该值必须是在 <filter> 元素中声明过的过滤器的名字。

<url-pattern> 设置 Filter 所拦截的请求路径（过滤器关联的 URL 样式）。

<filter-class> 指定完成过滤操作所使用的具体类。

<dispatcher> 指定过滤器所拦截的资源被 Servlet 容器调用的方式，可以是 REQUEST、INCLUDE、FORWARD 和 ERROR 之一，默认为 REQUEST。用户可以设置多个 <dispatcher> 子元素用来指定 Filter 对资源的多种调用方式进行拦截。代码如下：

```
<filter-mapping>
        <filter-name>testFilter</filter-name>
        <url-pattern>/index.jsp</url-pattern>
        <dispatcher>REQUEST</dispatcher>
        <dispatcher>FORWARD</dispatcher>
</filter-mapping>
```

专家讲解

<dispatcher> 子元素可以设置的值及其意义。

REQUEST：当用户直接访问页面时，Web 容器将会调用过滤器。如果目标资源是通过 RequestDispatcher 的 include()或 forward()方法访问时，那么该过滤器就不会被调用。

INCLUDE：如果目标资源是通过 RequestDispatcher 的 include()方法访问时，那么该过滤器将被调用。除此之外，该过滤器不会被调用。

FORWARD：如果目标资源是通过 RequestDispatcher 的 forward()方法访问时，那么该过滤器将被调用，除此之外，该过滤器不会被调用。

ERROR：如果目标资源是通过声明式异常处理机制调用的，那么该过滤器将被调用。除此之外，过滤器不会被调用。

12.2 监听器(Listener)

12.2.1 监听器介绍

监听器是一个专门用于对其他对象身上发生的事件或状态改变进行监听和相应处理的对象，当被监视的对象发生情况时，立即采取相应行动。监听器其实就是一个实现特定接口的普通 Java 程序，这个程序专门用于监听另一个 Java 对象的方法调用或属性改变，当被监听对象发生上述事件后，监听器某个方法立即被执行。

平时做开发的时候，是写监听器去监听其他对象。如果想设计一个对象，让这个对象可以被别的对象监听又该怎么做呢，可以按照严格的事件处理模型来设计一个对象，这个对象就可以被别的对象监听。事件处理模型涉及三个组件：事件源、事件对象、事件监听器。

下面来按照事件处理模型来设计一个 Person 对象，具体代码如下。

```
package com.isoft.listener;
//设计一个 Person 类作为事件源，这个类的对象的行为（比如吃饭、跑步）可以被其他的
对象监听
public class Person {
    //在 Person 类中定义一个 PersonListener 变量来记住传递进来的监听器
    private PersonListener listener;
    //设计 Person 的行为：吃
    public void eat() {
            if(listener != null) {
                    listener.doeat(new Event(this));
            }
    }
    //设计 Person 的行为：跑
    public void run() {
            if(listener != null) {
                    listener.dorun(new Event(this));
            }
    }
    //这个方法是用来注册对 Person 类对象的行为进行监听的监听器
    public void registerListener(PersonListener listener) {
            this.listener = listener;
    }
}
//设计 Person 类（事件源）的监听器接口
interface PersonListener {
    void doeat(Event e);
    void dorun(Event e);
}
//设计事件类，用来封装事件源
class Event {
    private Person source;
    public Event() {
    }
    public Event(Person source) {
            this.source = source;
```

```
    }
    public Person getSource() {
        return source;
    }
    public void setSource(Person source) {
        this.source = source;
    }
}
```

经过这样的设计，Person 类的对象就可以被其他对象监听了。测试代码如下。

```
public class TestPerson {
    public static void main(String[] args) {
        Person p = new Person();
        //注册监听 p 对象行为的监听器
        p.registerListener(new PersonListener() {
            //监听 p 吃东西这个行为
            public void doeat(Event e) {
                Person p = e.getSource();
                System.out.println(p + " 在吃东西 ");
            }
            //监听 p 跑步这个行为
            public void dorun(Event e) {
                Person p = e.getSource();
                System.out.println(p + " 在跑步 ");
            }
        });
        p.eat();
        p.run();
    }
}
```

运行结果如图 12.5 所示。

```
com.isoft.listener.Person@659e0bfd在吃东西
com.isoft.listener.Person@659e0bfd在跑步
```

图 12.5　运行结果

12.2.2 JavaWeb 中的监听器

JavaWeb 中的监听器是 Servlet 规范中定义的一种特殊类，它用于监听 Web 应用程序中的 ServletContext、HttpSession 和 ServletRequest 等域对象的创建与销毁事件以及监听这些域对象中的属性发生修改的事件。

在 Servlet 规范中定义了多种类型的监听器，它们用于监听的事件源分别为 ServletContext、HttpSession 和 ServletRequest 这三个域对象。

Servlet 规范针对这三个对象上的操作，又把多种类型的监听器划分为以下三种类型。

（1）监听域对象自身创建和销毁的事件监听器。

（2）监听域对象中属性的增加和删除的事件监听器。

（3）监听绑定到 HttpSession 域中的某个对象的状态的事件监听器。

12.2.2.1 监听 ServletContext 域对象的创建和销毁

ServletContextListener 接口用于监听 ServletContext 对象的创建和销毁事件。实现了 ServletContextListener 接口的类都可以对 ServletContext 对象的创建和销毁进行监听。

（1）当 ServletContext 对象被创建时，激发 contextInitialized（ServletContextEvent sce）方法。

（2）当 ServletContext 对象被销毁时，激发 contextDestroyed（ServletContextEvent sce）方法。

ServletContext 域对象创建和销毁时机分别如下。

（1）创建：服务器启动针对每一个 Web 应用创建 ServletContext。

（2）销毁：服务器关闭前先关闭代表每一个 Web 应用的 ServletContext。

案例 12.4：编写一个 MyServletContextListener 类，实现 ServletContextListener 接口，监听 ServletContext 对象的创建和销毁

1. 编写监听器

代码如下。

```
@WebListener
public class MyServletContextListener implements ServletContextListener {
   public void contextDestroyed(ServletContextEvent arg0) {
     System.out.println("ServletContext 对象销毁 ");
   }
   public void contextInitialized(ServletContextEvent arg0) {
    System.out.println("ServletContext 对象创建 ");
   }
}
```

2. 在 web.xml 文件中注册监听器

要想监听事件源，必须将监听器注册到事件源上才能够实现对事件源的行为动作进行监听。在 JavaWeb 中，监听的注册是在 web.xml 文件中进行配置的，代码如下。

```
<!-- 注册针对 ServletContext 对象进行监听的监听器 -->
<listener>
<description>ServletContextListener 监听器 </description>
<listener-class>com.isoft.web.listener.MyServletContextListener</listener-class>
</listener>
```

经过以上两个步骤，就完成了监听器的编写和注册，Web 服务器在启动时，MyServletContextListener 监听器就可以对 ServletContext 对象进行监听了。

12.2.2.2 监听 HttpSession 域对象的创建和销毁

HttpSessionListener 接口用于监听 HttpSession 对象的创建和销毁。

（1）创建一个 session 时，激发 sessionCreated（HttpSessionEvent se）方法。

（2）销毁一个 session 时，激发 sessionDestroyed（HttpSessionEvent se）方法。

实例 12.5：编写一个 MyHttpSessionListener 类，实现 HttpSessionListener 接口，监听 HttpSession 对象的创建和销毁

1. 编写监听器

代码如下。

```
//@WebListener
public class MyHttpSessionListener implements HttpSessionListener {
    public void sessionCreated(HttpSessionEvent arg0) {
            System.out.println(arg0.getSession() + " 创建了!! ");
    }
    public void sessionDestroyed(HttpSessionEvent arg0) {
            System.out.println("session 销毁了!! ");
    }
}
```

2. 在 web.xml 文件中注册监听器

代码如下。

```
<!-- 注册针对 HttpSession 对象进行监听的监听器 -->
<listener>
<description>HttpSessionListener 监听器 </description>
<listener-class>com.isoft. listener. MyHttpSessionListener</listener-class>
</listener>
<!-- 配置 HttpSession 对象的销毁时机 -->
<session-config>
<!-- 配置 HttpSession 对象在 1 分钟之后销毁 -->
<session-timeout>1</session-timeout>
</session-config>
```

当访问 JSP 页面时，HttpSession 对象就会创建，此时可以在 HttpSessionListener 观察到 HttpSession 对象的创建过程，也可以写一个 JSP 页面观察 HttpSession 对象创建的过程。

实现步骤如下。

```
index.jsp
<body>
```

当 JSP 页面被创建时，HttpSession 会被创建，创建好的 session 的 id 是：

```
{pageContext.session.id}
</body>
```

运行结果如图 12.6 所示。

http://localhost:8081/Chart13_FilterAndListener/index.jsp

一访问JSP页面，HttpSession就创建了，创建好的Session的id是：D1BF8EA6D9942B94F6AD65DFBAFDA089

图 12.6 运行结果

12.2.2.3 监听 ServletRequest 域对象的创建和销毁

ServletRequestListener 接口用于监听 ServletRequest 对象的创建和销毁。

（1）Request 对象被创建时，监听器的 requestInitialized（ServletRequestEvent sre）方法将会被调用。

（2）Request 对象被销毁时，监听器的 requestDestroyed（ServletRequestEvent sre）方法将会被调用。

ServletRequest 域对象创建和销毁时机分别如下。

（1）创建：用户每一次访问都会创建 Request 对象。

（2）销毁：当前访问结束，Request 对象就会销毁。

实例 12.6：编写一个 MyServletRequestListener 类，实现 ServletRequestListener 接口，监听 ServletRequest 对象的创建和销毁

代码如下。

```
@WebListener
public class MyServletRequestListener implements ServletRequestListener {
    public void requestDestroyed(ServletRequestEvent arg0) {
            System.out.println(arg0.getServletRequest() + " 销毁了!! ");
    }
    public void requestInitialized(ServletRequestEvent arg0) {
            System.out.println(arg0.getServletRequest() + " 创建了!! ");
    }
}
```

用户每一次访问都会创建 Request 对象，当访问结束后，Request 对象就会销毁。以上就是对监听器的一些简单讲解。

12.3 监听器的应用

12.3.1 监听域对象中属性变更的监听器

域对象中属性变更的事件监听器就是用来监听 ServletContext、HttpSession、HttpServletRequest 这三个对象中的属性变更信息事件的监听器。

这三个监听器接口分别是 ServletContextAttributeListener、HttpSessionAttributeListener 和 ServletRequestAttributeListener，这三个接口中都定义了三个方法来处理被监听对象中的属性的增加、删除和替换的事件，同一个事件在这三个接口中对应的方法名称完全相同，只是接受的参数类型不同。

12.3.1.1 attributeAdded 方法

当向被监听对象中增加一个属性时，Web 容器就调用事件监听器的 attributeAdded 方法进行响应，这个方法接收一个事件类型的参数，监听器可以通过这个参数来获得正在增加属性的域对象和被保存到域中的属性对象。

各个域属性监听器中的完整语法定义如下。

```
public void attributeAdded(ServletContextAttributeEvent scae)
public void attributeReplaced(HttpSessionBindingEvent  hsbe)
public void attributeRmoved(ServletRequestAttributeEvent srae)
```

12.3.1.2 attributeRemoved 方法

当删除被监听对象中的一个属性时，Web 容器调用事件监听器的 attributeRemoved 方法进行响应。

各个域属性监听器中的完整语法定义如下。

```
public void attributeRemoved(ServletContextAttributeEvent scae)
public void attributeRemoved(HttpSessionBindingEvent  hsbe)
public void attributeRemoved(ServletRequestAttributeEvent srae)
```

12.3.1.3 attributeReplaced 方法

当监听器的域对象中的某个属性被替换时，Web 容器调用事件监听器的 attributeReplaced 方法进行响应。

各个域属性监听器中的完整语法定义如下。

```
public void attributeReplaced(ServletContextAttributeEvent scae)
public void attributeReplaced(HttpSessionBindingEvent  hsbe)
public void attributeReplaced(ServletRequestAttributeEvent srae)
```

12.3.1.4 ServletContextAttributeListener 监听器案例。

编写 ServletContextAttributeListener 监听器监听 ServletContext 域对象的属性值变化情况，代码如下。

```
@WebListener
public class MyServletContextAttributeListener implements ServletContextAttributeListen-
er {
    @Override
    public void attributeAdded(ServletContextAttributeEvent scab) {
        String str = MessageFormat.format("ServletContext 域对象中添加了属
性 :{0}，属性值是 :{1}", scab.getName(), scab.getValue());
        System.out.println(str);
    }
    @Override
    public void attributeRemoved(ServletContextAttributeEvent scab) {
        String str = MessageFormat.format("ServletContext 域对象中删除属性 :{0},
属性值是 :{1}", scab.getName(), scab.getValue());
        System.out.println(str);
    }
    @Override
    public void attributeReplaced(ServletContextAttributeEvent scab) {
        String str = MessageFormat.format("ServletContext 域对象中替换了属性 :{0}
的值 ", scab.getName());
        System.out.println(str);
    }
}
```

编写 ServletContextAttributeListenerTest.jsp 测试页面，代码如下。

```
<body>
    <%
        // 往 application 域对象中添加属性
        application.setAttribute("name", " 融创软通 IT 学院 ");
        // 替换 application 域对象中 name 属性的值
        application.setAttribute("name", "rtct");
        // 移除 application 域对象中 name 属性
        application.removeAttribute("name");
    %>
</body>
```

运行结果如图 12.7 所示。

```
ServletContext域对象中添加了属性:org.apache.jasper.runtime.JspApplicationContextImpl,
ServletContext域对象中添加了属性:org.apache.jasper.compiler.ELInterpreter，属性值是:org.a
ServletContext域对象中添加了属性:name，属性值是:融创软通IT学院
ServletContext域对象中替换了属性:name的值
ServletContext域对象中删除属性:name，属性值是:rtct
```

图 12.7　运行结果

从运行结果中可以看到，ServletContextListener 监听器成功监听到了 ServletContext 域对象（application）中的属性值的变化情况。

12.3.1.5　ServletRequestAttributeListener 和 HttpSessionAttributeListener 监听器案例

编写监听器监听 HttpSession 和 HttpServletRequest 域对象的属性值变化情况，代码如下。

```
@WebListener
public class MyRequestAndSessionAttributeListener
        implements ServletRequestAttributeListener, HttpSessionAttributeListener {
    public void attributeRemoved(ServletRequestAttributeEvent arg0) {
        String str = MessageFormat.format("ServletRequest 域对象中删除属性 :{0}，属性值是 :{1}", arg0.getName(), arg0.getValue());
        System.out.println(str);
    }
    public void attributeAdded(ServletRequestAttributeEvent arg0) {
        String str = MessageFormat.format("ServletRequest 域对象中添加了属性 :{0}，属性值是 :{1}", arg0.getName(), arg0.getValue());
        System.out.println(str);
    }
    public void attributeReplaced(ServletRequestAttributeEvent arg0) {
        String str = MessageFormat.format("ServletRequest 域对象中替换了属性 :{0}的值 ", arg0.getName());
        System.out.println(str);
    }
    public void attributeAdded(HttpSessionBindingEvent arg0) {
        String str = MessageFormat.format("HttpSession 域对象中添加了属性 :{0}，属性值是 :{1}", arg0.getName(), arg0.getValue());
        System.out.println(str);
    }
    public void attributeRemoved(HttpSessionBindingEvent arg0) {
        String str = MessageFormat.format("HttpSession 域对象中删除属性 :{0}，属性值是 :{1}", arg0.getName(), arg0.getValue());
```

```
            System.out.println(str);
        }
        public void attributeReplaced(HttpSessionBindingEvent arg0) {
            String str = MessageFormat.format("HttpSession 域对象中替换了属性 :{0} 的值 ", arg0.getName());
            System.out.println(str);
        }
    }
```

编写 RequestAndSessionAttributeListenerTest.jsp 测试页面，代码如下。

```
<%
        // 往 session 域对象中添加属性
        session.setAttribute("uname", " 融创软通 IT 学院 ");
        // 替换 session 域对象中 uname 属性的值
        session.setAttribute("uname", " 教育培训事业部 ");
        // 移除 session 域对象中 uname 属性
        session.removeAttribute("uname");
        // 往 request 域对象中添加属性
        request.setAttribute("uname", " 融创软通 IT 学院 ");
        // 替换 request 域对象中 uname 属性的值
        request.setAttribute("uname", " 教育培训事业部 ");
        // 移除 request 域对象中 uname 属性
        request.removeAttribute("uname");
%>
```

运行结果如图 12.8 所示。

```
HttpSession域对象中添加了属性:uname，属性值是:融创软通IT学院
HttpSession域对象中替换了属性:uname的值
HttpSession域对象中删除属性:uname，属性值是:教育培训事业部
ServletRequest域对象中添加了属性:uname，属性值是:融创软通IT学院
ServletRequest域对象中替换了属性:uname的值
ServletRequest域对象中删除属性:uname，属性值是:教育培训事业部
```

图 12.8　运行结果

从运行结果中可以看到，HttpSessionAttributeListener 监听器和 ServletRequestAttributeListener 成功监听到了 HttpSession 域对象和 HttpServletRequest 域对象的属性值变化情况。

12.3.2　感知 session 绑定的事件监听器

保存在 session 域中的对象可以有多种状态：绑定（session.setAttribute("bean",Object)）到

session 中；从 session 域中解除（session.removeAttribute（"bean"））绑定；随 session 对象持久化到一个存储设备中；随 session 对象从一个存储设备中恢复。

Servlet 规范中定义了两个特殊的监听器接口“HttpSessionBindingListener”和“HttpSessionActivationListener”来帮助 JavaBean 对象了解自己在 session 域中的这些状态，实现这两个接口的类不需要在 web.xml 文件中进行注册。

12.3.2.1 HttpSessionBindingListener 接口

实现了 HttpSessionBindingListener 接口的 JavaBean 对象可以感知自己被绑定到 session 中和 session 中删除的事件。

（1）当对象被绑定到 HttpSession 对象中时，Web 服务器调用该对象的 void valueBound（HttpSessionBindingEvent event）方法。

（2）当对象从 HttpSession 对象中解除绑定时，Web 服务器调用该对象的 void valueUnbound（HttpSessionBindingEvent event）方法。

实例 12.7：HttpSessionBindingListener 的使用

```
@WebListener
public class JavaBeanDemo1 implements HttpSessionBindingListener {
    private String name;
    public JavaBeanDemo1（String name）{
            super（）;
            this.name = name;
    }
    public String getName（）{
            return name;
    }
    public void setName（String name）{
            this.name = name;
    }
    public void valueBound（HttpSessionBindingEvent arg0）{
            System.out.println（name + " 被加到 session 中了 "）;
    }
    public void valueUnbound（HttpSessionBindingEvent arg0）{
            System.out.println（name + " 被 session 踢出来了 "）;
    }
}
```

上述的 JavaBeanDemo1 这个 JavaBean 实现了 HttpSessionBindingListener 接口，那么这个 JavaBean 对象可以感知自己被绑定到 session 中和从 session 中删除的这两个操作，测试代码如下。

```
<%
    // 将 JavaBean 对象绑定到 session 中
    session.setAttribute("bean",new JavaBeanDemo1(" 融创软通 IT 学院 "));
    // 从 session 中删除 JavaBean 对象
    session.removeAttribute("bean");
%>
```

12.3.2.2 HttpSessionActivationListener 接口

实现了 HttpSessionActivationListener 接口的 JavaBean 对象可以感知自己被活化（反序列化）和钝化（序列化）的事件。

（1）当绑定到 HttpSession 对象中的 JavaBean 对象将要随 HttpSession 对象被钝化（序列化）之前，Web 服务器调用该 JavaBean 对象的 void sessionWillPassivate（HttpSessionEvent event）方法。这样 JavaBean 对象就可以知道自己将要和 HttpSession 对象一起被序列化（钝化）到硬盘中。

（2）当绑定到 HttpSession 对象中的 JavaBean 对象，随 HttpSession 对象被活化（反序列化）后，Web 服务器调用该 JavaBean 对象的 void sessionDidActive（HttpSessionEvent event）方法。这样 JavaBean 对象就可以知道自己将要和 HttpSession 对象一起被反序列化（活化）回到内存中。

JavaWeb 开发技术中的监听器技术在平时工作中的 JavaWeb 项目开发中用得是比较多的，因此必须掌握这门技术。

12.4 过滤器(Filter)常见应用

12.4.1 统一全站字符编码

通过配置参数 charset 指明使用何种字符编码，以处理 Html Form 请求参数的中文问题。

```
@WebFilter("/*")
public class CharacterEncodingFilter implements Filter {
    private FilterConfig filterConfig = null;
    private String defaultCharset = "UTF-8";
    public void doFilter(ServletRequest req, ServletResponse resp, FilterChain chain)
                throws IOException, ServletException {
        HttpServletRequest request = (HttpServletRequest) req;
        HttpServletResponse response = (HttpServletResponse) resp;
        String charset = filterConfig.getInitParameter("charset");
        if(charset == null) {
```

```
                    charset = defaultCharset;
            }
            request.setCharacterEncoding(charset);
            response.setCharacterEncoding(charset);
            response.setContentType("text/html;charset=" + charset);
            chain.doFilter(request, response);
    }
    public void init(FilterConfig filterConfig) throws ServletException {
            // 得到过滤器的初始化配置信息
            this.filterConfig = filterConfig;
    }
    @Override
    public void destroy() {
    }
}
```

web.xml 文件中的配置如下。

```
    <filter>
            <filter-name>CharacterEncodingFilter</filter-name>
            <filter-class>com.isoft.filter.CharacterEncodingFilter</filter-class>
            <init-param>
                    <param-name>charset</param-name>
                    <param-value>UTF-8</param-value>
            </init-param>
    </filter>
    <filter-mapping>
            <filter-name>CharacterEncodingFilter</filter-name>
            <url-pattern>/*</url-pattern>
    </filter-mapping>
```

12.4.2 实现用户自动登录

实现用户自动登录的思路如下。

(1)在用户登录成功后，发送一个名称为 user 的 cookie 给客户端，cookie 的值为用户名和 md5 加密后的密码。

(2)编写一个 AutoLoginFilter，这个 Filter 检查用户是否带有名称为 user 的 cookie，如果有，则调用 dao 查询 cookie 的用户名和密码是否和数据库匹配，匹配则向 session 中存入 user 对象(即用户登录标记)，以实现程序的自动登录。

核心代码如下。

```
处理用户登录的控制器：LoginServlet
@WebServlet("/loginServlet")
public class LoginServlet extends HttpServlet {
    private static final long serialVersionUID = 1L;
    public void doGet(HttpServletRequest request, HttpServletResponse response) throws
ServletException, IOException {
            String username = request.getParameter("username");
            String password = request.getParameter("password");
            UserDao dao = new UserDao();
            User user = dao.find(username, password);
            if(user == null) {
                request.setAttribute("message", "用户名或密码不对!! ");
                request.getRequestDispatcher("/message.jsp").forward(request, response);
                return;
            }
            request.getSession().setAttribute("user", user);
            // 发送自动登录 cookie 给客户端浏览器进行存储
            sendAutoLoginCookie(request, response, user);
            request.getRequestDispatcher("/index.jsp").forward(request, response);
    }
    private void sendAutoLoginCookie(HttpServletRequest request, HttpServletResponse re-
sponse, User user) {
            if(request.getParameter("logintime") != null) {
                int logintime = Integer.parseInt(request.getParameter("logintime"));
                // 创建 cookie,cookie 的名字是 autologin，值是用户登录的用户名和
密码，用户名和密码之间使用“.”进行分割，密码经过 md5 加密处理
                Cookie cookie = new Cookie("autologin", user.getUsername() + "." +
WebUtils.md5(user.getPassword()));
                // 设置 cookie 的有效期
                cookie.setMaxAge(logintime);
                // 设置 cookie 的有效路径
                cookie.setPath(request.getContextPath());
                // 将 cookie 写入到客户端浏览器
                response.addCookie(cookie);
```

```
            }
        }
        public void doPost(HttpServletRequest request, HttpServletResponse response) throws
ServletException, IOException {
            doGet(request, response);
        }
    }
```

处理用户自动登录的过滤器：AutoLoginFilter

```
    @WebFilter("/*")
    public class AutoLoginFilter implements Filter {
        public void doFilter(ServletRequest req, ServletResponse res, FilterChain chain)
                    throws IOException, ServletException {
            HttpServletRequest request = (HttpServletRequest)req;
            HttpServletResponse response = (HttpServletResponse) res;
            // 如果已经登录了，就直接 chain.doFilter(request, response)放行
            if(request.getSession().getAttribute("user") != null) {
                    chain.doFilter(request, response);
                    return;
            }
            // 1. 得到用户带过来的 authlogin 的 cookie
            String value = null;
            Cookie cookies[] = request.getCookies();
            for(int i = 0; cookies != null && i < cookies.length; i++) {
                    if(cookies[i].getName().equals("autologin")) {
                            value = cookies[i].getValue();
                    }
            }
            // 2. 得到 cookie 中的用户名和密码
            if(value != null) {
                    String username = value.split("\\.")[0];
                    String password = value.split("\\.")[1];
                    // 3. 调用 dao 获取用户对应的密码
                    UserDao dao = new UserDao();
                    User user = dao.find(username);
                    String dbpassword = user.getPassword();
                    // 4. 检查用户带过来的 md5 的密码和数据库中的密码是否匹配，如
匹配则自动登录
```

```
                if(password.equals(WebUtils.md5(dbpassword))) {
                    request.getSession().setAttribute("user", user);
                }
            }
            chain.doFilter(request, response);
        }
    }
```

如果想取消自动登录，那么可以在用户注销时删除自动登录 cookie，核心代码如下：

```
@WebServlet("/cancelAutoLoginServlet")
public class CancelAutoLoginServlet extends HttpServlet {
    public void doGet(HttpServletRequest request, HttpServletResponse response) throws
ServletException, IOException {
            // 移除存储在 session 中的 user
            request.getSession().removeAttribute("user");
            // 移除自动登录的 cookie
            removeAutoLoginCookie(request, response);
            // 注销用户后跳转到登录页面
            request.getRequestDispatcher("/login.jsp").forward(request, response);
    }
    private void removeAutoLoginCookie(HttpServletRequest request, HttpServletResponse
response) {
            // 创建一个名字为 autologin 的 cookie
            Cookie cookie = new Cookie("autologin", "");
            // 将 cookie 的有效期设置为 0，命令浏览器删除该 cookie
            cookie.setMaxAge(0);
            // 设置要删除的 cookie 的 path
            cookie.setPath(request.getContextPath());
            response.addCookie(cookie);
    }
}
```

12.5 监听器(Listener)在开发中的应用

监听器在 JavaWeb 开发中用得比较多，下面介绍监听器(Listener)在开发中的常见应用。

在 JavaWeb 应用开发中，有时候需要统计当前在线的用户数，此时就可以使用监听器技术来实现了。

案例 12.8:统计当前在线人数

```
@WebListener
public class OnLineCountListener implements HttpSessionListener {
    @Override
    public void sessionCreated(HttpSessionEvent se) {
            ServletContext context = se.getSession().getServletContext();
            Integer onLineCount = (Integer) context.getAttribute("onLineCount");
            if(onLineCount == null) {
                    context.setAttribute("onLineCount", 1);
            } else {
                    onLineCount++;
                    context.setAttribute("onLineCount", onLineCount);
            }
    }
    @Override
    public void sessionDestroyed(HttpSessionEvent se) {
            ServletContext context = se.getSession().getServletContext();
            Integer onLineCount = (Integer) context.getAttribute("onLineCount");
            if(onLineCount == null) {
                    context.setAttribute("onLineCount", 1);
            } else {
                    onLineCount--;
                    context.setAttribute("onLineCount", onLineCount);
            }
    }
}
```

小结

Servlet 过滤器是 Servlet 程序的一种特殊用法,主要用来完成一些通用的操作,如编码的过滤、判断用户的登录状态。过滤器使得 Servlet 开发者能够在客户端请求到达 Servlet 资源之前被截获,在处理之后再发送给被请求的 Servlet 资源,并且还可以截获响应,修改之后再发送给用户。而 Servlet 监听器可以监听客户端发出的请求、服务器端的操作,通过监听器可以自动激发一些操作,如监听在线人数。

经典面试题

1. 什么是过滤器？
2. 如何创建过滤器 Filter?
3. 什么是过滤器链？
4. 描述一下 Filter 的声明周期。
5. 过滤器的注解配置是什么？
6. 什么是监听器？
7. 监听器的分类有哪些？
8. 监听器主要应用于什么场景？
9. 监听器（Listener）的工作原理是什么？
10. 如何使用监听器实现统计在线人数？

跟我上机

1. 使用 Filter 实现未登录用户自动跳转到登录页面。
2. 使用 Filter 控制在线玩游戏的时间。
3. 统计在线用户数以及同账号登录踢出另一用户。

第 13 章　MVC 开发模式

本章要点(学会后请在方框里打钩):

- □ 了解什么是 MVC 开发模式
- □ 了解什么是 Model Ⅰ 模式
- □ 掌握什么是 Model Ⅱ 模式
- □ 掌握 Model Ⅱ 模式的优缺点
- □ 掌握如何使用 MVC 模式开发项目

SUN 公司推出 JSP 技术后，同时也推荐了两种 Web 应用程序的开发模式：一种是 JSP+JavaBean 模式（Model Ⅰ），另一种是 Servlet+JSP+JavaBean 模式（Model Ⅱ）。

13.1 Model Ⅰ 模式

13.1.1 JSP+JavaBean 开发模式架构

在 JSP+JavaBean 架构中，JSP 负责控制逻辑、表现逻辑、业务对象（JavaBean）的调用。

JSP+JavaBean 模式适合开发业务逻辑不太复杂的 Web 应用程序，在这种模式下，JavaBean 用于封装业务数据，JSP 既负责处理用户请求，又显示数据。

13.1.2 使用 Model Ⅰ 开发模式编写简单计算器

首先分析一下 JSP 和 JavaBean 各自的职责：JSP 负责显示计算器（Calculator）页面，供用户输入计算数据，并显示计算后的结果；JavaBean 负责接收用户输入的计算数据并进行计算，JavaBean 具有 firstNum、secondNum、result、operator 属性，并提供一个 Calculate 方法。

13.1.2.1 编写 CalculatorBean，负责接收用户输入的计算数据并进行计算

CalculatorBean 代码如下。

```
package com.isoft.bean;
import java.math.BigDecimal;
public class CalculatorBean {
    private double firstNum;
    private double secondNum;
    private char operator = '+';
    private double result;
    public double getFirstNum() {
            return firstNum;
    }
    public void setFirstNum(double firstNum) {
            this.firstNum = firstNum;
    }
    public double getSecondNum() {
            return secondNum;
    }
    public void setSecondNum(double secondNum) {
            this.secondNum = secondNum;
    }
```

```
    public char getOperator() {
            return operator;
    }
    public void setOperator(char operator) {
            this.operator = operator;
    }
    public double getResult() {
            return result;
    }
    public void setResult(double result) {
            this.result = result;
    }
    public void calculate() {
            switch (this.operator) {
            case '+': {
                    this.result = this.firstNum + this.secondNum;
                    break;
            }
            case '-': {
                    this.result = this.firstNum - this.secondNum;
                    break;
            }
            case '*': {
                    this.result = this.firstNum * this.secondNum;
                    break;
            }
            case '/': {
                    if(this.secondNum == 0) {
                            throw new RuntimeException("被除数不能为 0!!!");
                    }
                    this.result = this.firstNum / this.secondNum;
                    //四舍五入
                    this.result = new BigDecimal(this.result).setScale(2, BigDecimal.
ROUND_HALF_UP).doubleValue();
                    break;
            }
            default:
```

```
                throw new RuntimeException(" 对不起,传入的运算符非法! ");
            }
        }
    }
```

13.1.2.2 编写 calculator.jsp,负责显示计算器(Calculator)页面,供用户输入计算数据,并显示计算后的结果

calculator.jsp 页面代码如下。

```
<%@ page language="java" import="java.util.*" pageEncoding="UTF-8"%>
<%-- 使用 com.isoft.domain.CalculatorBean --%>
<jsp:useBean id="calcBean" class="com.isoft.bean.CalculatorBean" />
<jsp:setProperty name="calcBean" property="*" />
<%
    calcBean.calculate();
%>
<!DOCTYPE HTML>
<html>
<head>
<title> 使用【jsp+javabean 开发模式】开发的简单计算器 </title>
</head>
<body>
    <br /> 计算结果是:
    <jsp:getProperty name="calcBean" property="firstNum" />
    <jsp:getProperty name="calcBean" property="operator" />
    <jsp:getProperty name="calcBean" property="secondNum" />
    =
    <jsp:getProperty name="calcBean" property="result" />
    <br />
    <hr>
    <br />
    <form action="/calculator.jsp" method="post">
            <table border="1px">
                    <tr>
                            <td colspan="2"> 简单的计算器 </td>
                    </tr>
                    <tr>
                            <td> 第一个参数 </td>
```

```
                    <td><input type="text" name="firstNum"></td>
                </tr>
                <tr>
                    <td> 运算符 </td>
                    <td><select name="operator">
                            <option value="+">+</option>
                            <option value="-">-</option>
                            <option value="*">*</option>
                            <option value="/">/</option>
                    </select></td>
                </tr>
                <tr>
                    <td> 第二个参数 </td>
                    <td><input type="text" name="secondNum"></td>
                </tr>
                <tr>
                    <td colspan="2"><input type="submit" value=" 计算 "></td>
                </tr>
            </table>
    </form>
</body>
</html>
```

运行结果如图 13.1 所示。

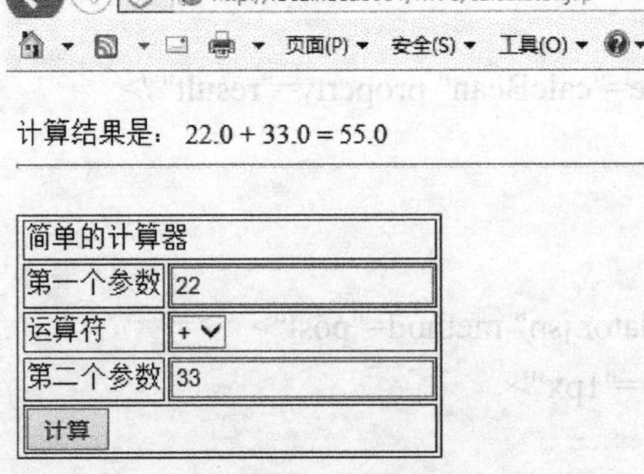

图 13.1　运行结果

13.2 Model Ⅱ模式

平时的 JavaWeb 项目开发中，在不使用第三方 MVC（Model-View-Controller，模型 - 视图 - 控制器）开发框架的情况下，通常会选择 Servlet+JSP+JavaBean 开发模式来开发 JavaWeb 项目，它适合开发复杂的 Web 应用，在这种模式下，Servlet 负责处理用户请求，JSP 负责数据显示，JavaBean 负责封装数据。Servlet+JSP+JavaBean 模式程序各个模块之间层次清晰，Web 开发推荐采用此种模式。

在讲解 Servlet+JSP+JavaBean 开发模式之前，先简单了解一下 MVC 开发模式。

13.2.1 标准 MVC 模型概述

MVC 模型是一种架构型的模式，本身不引入新功能，只是帮助我们让开发的结构组织更加合理，使展示与模型分离，流程控制逻辑、业务逻辑调用与展示逻辑分离，如图 13.2 所示。

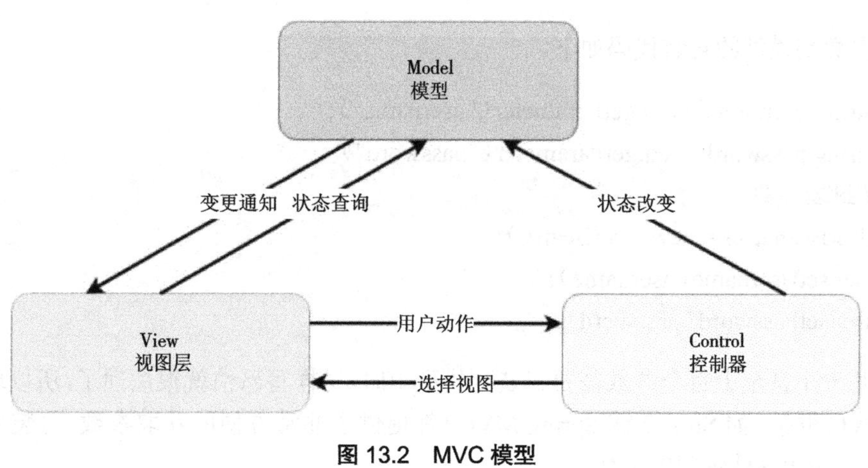

图 13.2 MVC 模型

13.2.2 MVC 的概念

Model（模型）：数据模型，提供要展示的数据，因此包含数据和行为，可以认为是领域模型（Domain）或 JavaBean 组件（包含数据和行为），不过现在一般都分离开来，即 Value Object（数据）和服务层（行为）。也就是模型提供了模型数据查询和模型数据的状态更新等功能，包括数据和业务。

View（视图）：负责进行模型的展示，一般就是我们见到的用户界面，客户想看到的内容。

Controller（控制器）：接收用户请求，委托给模型进行处理（状态改变），处理完毕后把返回的模型数据返回给视图，由视图负责展示，也就是说控制器做了调度员的工作。

13.2.3 Servlet+JSP+JavaBean 开发模式

Servlet+JSP+JavaBean 架构其实可以认为就是我们所说的 Web MVC 模型，只是控制器采用 Servlet，模型采用 JavaBean，视图采用 JSP。

13.3 Model Ⅱ开发模式的缺点

Servlet+JSP+JavaBean（Web MVC）架构虽然实现了视图和模型分离以及控制逻辑和展示逻辑分离，但也有很多的缺点。

13.3.1 Servlet 作为控制器的缺点

Model Ⅱ开发模式的控制器使用 Servlet，使用 Servlet 作为控制器有以下几个缺点。

（1）因为控制逻辑可能会比较复杂，所以如果能按照规约编写能有效减少复杂度，例如，请求参数 submitFlag=toLogin，可以改为直接调用 toLogin 方法来简化控制逻辑，另外由于每个模块都需要一个控制器，从而使得控制器逻辑更加复杂。现在流行的 WebMVC 框架（例如 Struts2）均支持直接调用法，这样的处理机制在 Struts2 中被称为“动态方法调用”。

（2）请求参数到模型的封装比较麻烦，如果能交给框架来做这件事情，可以从中得到解放。

请求参数到模型的封装代码如下。

```
String username = req.getParameter("username");
String password = req.getParameter("password");
// 封装参数
UserBean user = new UserBean();
user.setUsername(username);
user.setPassword(password);
```

当有几十个甚至上百个参数需要封装到模型中时，这样写恐怕就很烦琐了，所以现在流行的 Web MVC 框架（如 Struts2 或 Spring MVC）都提供了非常方便的获取参数、封装参数到模型的机制，以减少这些烦琐的工作。

（3）选择下一个视图，严重依赖 Servlet API，这样很难或基本不可能更换视图。

例如：使用 Servlet API 提供的 request 对象的 getRequestDispatcher 方法选择要展示给用户看的视图。

```
    private void toLogin(HttpServletRequest req, HttpServletResponse resp) throws Servlet-
Exception, IOException {
            // 使用 Servlet API 提供的 request 对象的 getRequestDispatcher 方法选择视图
            // 此处和 JSP 视图技术紧密耦合，更换其他视图技术几乎不可能
            request.getRequestDispatcher("/mvc/login.jsp").forward(request, response);
    }
```

（4）给视图传输需要展示的数据模型也需要使用 ServletAPI，当需要更换其他视图技术时，ServletAPI 的相关调用也需要同步更换，很麻烦。

例如：使用 Servlet API 提供的 request 对象向视图传输要展示的模型数据。

```
// 使用 Servlet API 提供的 request 对象给视图 login.jsp 传输要展示的模型数据(user)
request.setAttribute("user", user);
request.getRequestDispatcher("/mvc/login.jsp").forward(request, response)
```

13.3.2 JavaBean 作为模型的缺点

JavaBean 组件类既负责收集封装数据，又要进行业务逻辑处理，这样可能造成 JavaBean 组件类很庞大，所以现在的项目一般采用三层架构，而不直接采用 JavaBean。

13.3.3 JSP 作为视图的缺点

1. 控制逻辑可能比较复杂

其实可以按照规约，如请求参数 submitFlag=toLogin，直接调用 toLogin 方法来简化控制逻辑；而且基本每个模块都需要一个控制器，可能造成控制逻辑很复杂。

2. 请求参数到模型的封装比较麻烦

如果能将参数封装这件事交给框架来做，我们将不再需要自己编写参数封装语句，从而得到解放。

13.4 综合案例——Model Ⅱ模式开发用户登录注册

一个良好的 JavaWeb 项目架构应该具有很多个包，这样不仅显得层次分明，而且各个层之间的职责也很清晰明了。在搭建 JavaWeb 项目架构时，按照下面的顺序创建包：domain → dao → dao.impl → service → service.impl → web.controller → web.UI → web.filter → web.listener → util → junit.test，包的层次创建好了，项目的架构也就定下来了。当然，在实际的项目开发中，也不一定是完完全全按照上面说的来创建包的层次结构，而是根据项目的实际情况，可能还需要创建其他的包，应根据项目的需要来决定。

13.4.1 创建目录结构

创建目录结构，如图 13.3 所示。

```
Model II
  JAX-WS Web Services
  Java Resources
    src/main/java
      com.isoft.dao
        IUserDao.java
      com.isoft.dao.impl
        UserDaoImpl.java
          UserDaoImpl
      com.isoft.domain
        User.java
      com.isoft.exception
        UserExistException.java
      com.isoft.service
        IUserService.java
      com.isoft.service.impl
        UserServiceImpl.java
      com.isoft.util
        DBUtils.java
          DBUtils
        WebUtils.java
      com.isoft.web.controller
        LoginServlet.java
        LogoutServlet.java
        RegisterServlet.java
      com.isoft.web.formbean
        RegisterFormBean.java
      com.isoft.web.listener
      com.isoft.web.UI
        LoginUIServlet.java
        RegisterUIServlet.java
      junit.test
    src/main/resources
      jdbcConfig.properties
    src/test/java
    src/test/resources
    Libraries
      JRE System Library [JavaSE-1.8]
      Maven Dependencies
        standard-1.1.2.jar - C:\Users\zjj\.m2\rep
        commons-beanutils-1.9.3.jar - C:\Users\
        commons-logging-1.2.jar - C:\Users\zjj\
        commons-collections-3.2.2.jar - C:\User
        javax.servlet-api-4.0.0.jar - C:\Users\zjj\
        jstl-1.2.jar - C:\Users\zjj\.m2\repository\
        mysql-connector-java-5.1.40.jar - C:\Us
  JavaScript Resources
  Deployed Resources
```

图 13.3 创建目录结构

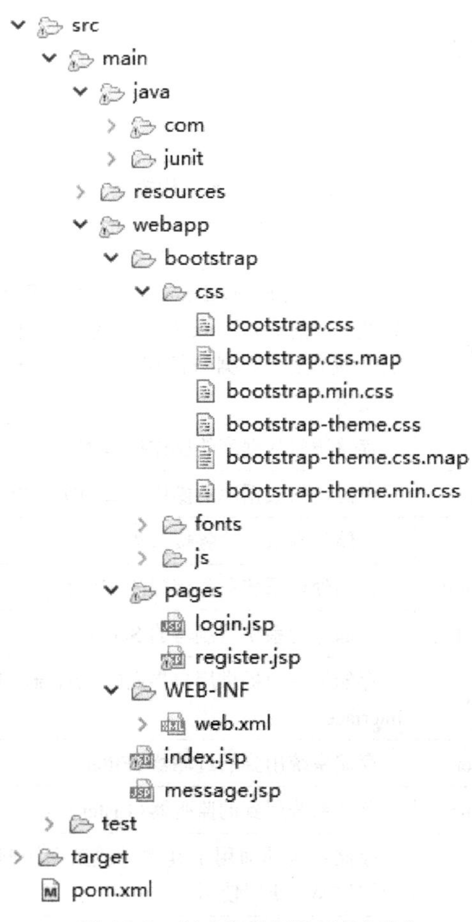

图 13.3 创建好的目录结构(续)

13.4.2 项目所需要的开发包(jar 包)

项目所需要的开发包(jar 包)见表 13.1。

表 13.1 项目所需要的开发包

序 号	开发包名称	描 述
1	mysql-connector-java-5.1.40.jar	MySQl 数据库驱动包
2	javax.servlet-api- 4.0.jar	Servlet-api 包
3	commons-beanutils-1.9.3.jar	工具类,用于处理 bean 对象
4	commons-logging.jar	commons-beanutils-1.9.3.jar 的依赖 jar 包
5	jstl.jar	JSTL 标签库和 EL 表达式依赖包
6	standard.jar	JSTL 标签库和 EL 表达式依赖包

13.4.3 项目目录结构描述

项目目录结构描述见表 13.2。

表 13.2 项目目录结构描述

序 号	包 名	描 述	所属层次
1	com.isoft.domain	存储系统的 JavaBean 类(只包含简单的属性以及属性对应的 get 和 set 方法,不包含具体的业务处理方法),提供给“数据访问层”“业务处理层”“Web 层”来使用	domain(域模型)层
2	com.isoft.dao	存储访问数据库的操作接口类	数据访问层
3	com.isoft.dao.impl	存储访问数据库的操作接口的实现类	
4	com.isoft.service	存储处理系统业务接口类	业务处理层
5	com.isoft.service.impl	存储处理系统业务接口的实现类	
6	com.isoft.web.controller	存储作为系统控制器的 Servlet	Web 层(表现层)
7	com.isoft.web.UI	存储为用户提供用户界面的 Servlet(UI 指的是 User Interface)	
8	com.isoft.web.filter	存储系统用到的过滤器(Filter)	
9	com.isoft.web.listener	存储系统用到的监听器(Listener)	
10	com.isoft.util	存储系统的通用工具类,供“数据访问层”“业务处理层”“Web 层”使用	
11	junit.test	存储系统的测试类	

13.4.4 项目源码

为了节省空间,本例去掉了所有导入包的代码。

pom.xml 代码如下。

```
<dependencies>
    <dependency>
            <groupId>taglibs</groupId>
            <artifactId>standard</artifactId>
            <version>1.1.2</version>
    </dependency>
            <dependency>
                <groupId>commons-beanutils</groupId>
                <artifactId>commons-beanutils</artifactId>
                <version>1.9.3</version>
```

```
            </dependency>
            <dependency>
                    <groupId>javax.servlet</groupId>
                    <artifactId>javax.servlet-api</artifactId>
                    <version>4.0.0</version>
            </dependency>
            <dependency>
                    <groupId>jstl</groupId>
                    <artifactId>jstl</artifactId>
                    <version>1.2</version>
            </dependency>
            <dependency>
                    <groupId>mysql</groupId>
                    <artifactId>mysql-connector-java</artifactId>
                    <version>5.1.40</version>
            </dependency>
    </dependencies>
```

index.jsp 代码如下。

```
<%@ page language="java" pageEncoding="UTF-8"%>
<%-- 为了避免在 JSP 页面中出现 Java 代码，这里引入 JSTL 标签库，利用 JSTL 标签库提供的标签来做一些逻辑判断处理 --%>
<%@ taglib uri="http://java.sun.com/jsp/jstl/core" prefix="c"%>
<!DOCTYPE HTML>
<html>
<head>
<title> 首页 </title>
<link href="${pageContext.request.contextPath}/bootstrap/css/bootstrap.min.css" rel="stylesheet" />
<script type="text/javascript">
    function doLogout() {
            // 访问 LogoutServlet 注销当前登录的用户
            window.location.href = "${pageContext.request.contextPath}/logoutServlet";
    }
</script>
</head>
<body>
```

```
        <h1>MVC Model II 模式案例 </h1>
        <hr />
        <c:if test="${user==null}">
                <a href="${pageContext.request.contextPath}/registerUIServlet" target="_blank"> 注册
</a>
                <a href="${pageContext.request.contextPath}/loginUIServlet"> 登录 </a>
        </c:if>
        <c:if test="${user!=null}">
                欢迎您：${user.uname}
                <input type="button" value=" 退出登录 " onclick="doLogout()">
        </c:if>
        <hr />
    </body>
    </html>
```

message.jsp 代码如下。

```
<%@ page language="java" pageEncoding="UTF-8"%>
<!DOCTYPE HTML>
<html>
<head>
<title> 全局消息显示页面 </title>
<link href="${pageContext.request.contextPath}/bootstrap/css/bootstrap.min.css" rel="stylesheet" />
</head>
<body>${message}
</body>
</html>
```

login.jsp 代码如下。

```
<%@ page language="java" pageEncoding="UTF-8"%>
<!DOCTYPE HTML>
<html>
<head>
<title> 用户登录 </title>
<link href="${pageContext.request.contextPath}/bootstrap/css/bootstrap.min.css" rel="stylesheet" />
<link href="${pageContext.request.contextPath}/bootstrap/css/bootstrap-theme.min.css" rel="stylesheet" />
```

```
    </head>
    <body>
        <div class="center-block" style="width: 40%">
                <div class="panel panel-primary">
                        <form action="${pageContext.request.contextPath}/loginServlet" method=
"post">
                                <div class="panel-heading">
                                        <h3> 用户登录 </h3>
                                </div>
                                <div class="panel-body">
                                        <div class="form-inline">
                                                <label class="h4" for="ds_host"> 用户名 </
label> <input
                                                        class="form-control" type="text" name=
"username"
                                                        style="width: 100%"><br />
                                        </div>
                                        <br>
                                        <div class="form-inline">
                                                <label class="h4" for="ds_name"> 密  
  码 </label> <input
                                                        class="form-control" type="password"
name=" password"
                                                        style="width: 100%"><br />
                                        </div>
                                        <br>
                                        <div align="center">
                                                <input class="btn btn-success" type="submit"
value=" 登录 "> <a
                                                        class="btn btn-warning"
                                                        href="${pageContext.request.context-
Path}/pages/register.jsp"> 返回注册 </a>
                                        </div>
                                </div>
                        </form>
                </div>
        </div>
```

```
</body>
</html>
```

register.jsp 代码如下。

```
<%@ page language="java" pageEncoding="UTF-8"%>
<!DOCTYPE HTML>
<html>
<head>
<title> 用户注册 </title>
<link href="${pageContext.request.contextPath}/bootstrap/css/bootstrap.min.css" rel="stylesheet" />
</head>
<body style="text-align: center;">
    <div class="center-block" style="width: 40%">
        <form action="${pageContext.request.contextPath}/registerServlet" method="post">
            <table class="table table-hover table-bordered table-striped">
                <caption>
                    <h3 align="center"> 用户信息注册 </h3>
                </caption>
                <tr>
                    <td> 用户名 </td>
                    <td>
                        <%-- 使用 EL 表达式 ${} 提取存储在
request 对象中的 formbean 对象中封装的表单数据（formbean.uname）以及错误提示消息
（formbean.errors.uname）--%>
                        <input class="form-control" type="text" name=
"userName"
                        value="${formbean.userName}">${formbean.
errors.userName}
                    </td>
                </tr>
                <tr>
                    <td> 密码 </td>
                    <td><input class="form-control" type="password" name=
"userPwd"
                        value="${formbean.userPwd}">${formbean.
errors.userPwd}</td>
```

```
                        </tr>
                        <tr>
                                <td> 确认密码 </td>
                                <td><input class="form-control" type="password"
                                        name="confirmPwd" value="${formbean.con-
firmPwd}">${formbean.errors.confirmPwd}
                                </td>
                        </tr>
                        <tr>
                                <td> 邮箱 </td>
                                <td><input class="form-control" type="email" name="e-
mail"
                                        value="${formbean.email}">${formbean.er-
rors.email}</td>
                        </tr>
                        <tr>
                                <td> 生日 </td>
                                <td><input class="form-control" type="date" name="birth-
day"
                                        value="${formbean.birthday}">${formbean.
errors.birthday}</td>
                        </tr>
                        <tr>
                                <td colspan="2"><input type="submit" value=" 注册 "
                                        class="btn  btn-success"><input  type="reset"
value=" 清空 "
                                        class="btn btn-info"> <a
                                        href="${pageContext.request.contextPath}/
pages/login.jsp"> 返回登录 </a></td>
                        </tr>
                </table>
        </form>
    </div>
  </body>
  </html>
```

IUserDao.java 代码如下。

```
package com.isoft.dao;
public interface IUserDao {
    List<Map<String, Object>> find(String userName, String userPwd);
    void add(User user);
    List<Map<String, Object>> find(String userName);
}
```

UserDaoImpl.java 代码如下。

```
package com.isoft.dao.impl;
public class UserDaoImpl implements IUserDao {
    @Override
    public List<Map<String, Object>> find(String userName, String userPwd) {
            try {
                    DBUtils.getConnection();
                    String sql = "select * from tb_user where uname=? and upwd=?";
                    List<Map<String, Object>> query = DBUtils.query(sql, userName, us-
erPwd);
                    return query;
            } catch (Exception e) {
                    // e.printStackTrace();
                    throw new RuntimeException(e);
            }
    }
    @SuppressWarnings("deprecation")
    @Override
    public void add(User user) {
            try {
                    DBUtils.getConnection();
                    String sql = "insert into tb_user(id,uname,upwd,email,birthday) val-
ues(?,?,?,?,?)";
                    DBUtils.update(sql, user.getId(),user.getUserName(),user.getUserP-
wd(),user.getEmail(),user.getBirthday());
            } catch (Exception e) {
                    throw new RuntimeException(e);
            }
    }
```

```
    @Override
    public List<Map<String, Object>> find(String userName) {
        try {
            DBUtils.getConnection();
            String sql = "select * from tb_user where uname=?";
            List<Map<String, Object>> query = DBUtils.query(sql, userName);
            return query;
        } catch(Exception e) {
            throw new RuntimeException(e);
        }
    }
}
```

User.java 代码如下。

```
package com.isoft.domain;
public class User implements Serializable {
    private static final long serialVersionUID = -4313782718477229465L;
    private String id;
    private String userName;
    private String userPwd;
    private String email;
    private Date birthday;
    public String getId() {
        return id;
    }
    public void setId(String id) {
        this.id = id;
    }
    public String getUserName() {
        return userName;
    }
    public void setUserName(String userName) {
        this.userName = userName;
    }
    public String getUserPwd() {
        return userPwd;
    }
```

```
    public void setUserPwd(String userPwd) {
            this.userPwd = userPwd;
    }
    public String getEmail() {
            return email;
    }
    public void setEmail(String email) {
            this.email = email;
    }
    public Date getBirthday() {
            return birthday;
    }
    public void setBirthday(Date birthday) {
            this.birthday = birthday;
    }
}
```

UserExistException.java 代码如下。

```
package com.isoft.exception;
public class UserExistException extends Exception {
    public UserExistException() {
            super();
    }
    public UserExistException(String message, Throwable cause) {
            super(message, cause);
    }
    public UserExistException(String message) {
            super(message);
    }
    public UserExistException(Throwable cause) {
            super(cause);
    }
}
```

IUserService.java 代码如下。

```
package com.isoft.service;
public interface IUserService {
```

```
    void registerUser(User user) throws UserExistException;
    List<Map<String, Object>> loginUser(String userName, String userPwd);
}
```

UserServiceImpl.java 代码如下。

```
package com.isoft.service.impl;
public class UserServiceImpl implements IUserService {
    private IUserDao userDao = new UserDaoImpl();
        @Override
    public void registerUser(User user) throws UserExistException {
        if(userDao.find(user.getUserName()).size()!=0) {
            //这里编译时抛异常的原因：是上一层程序处理这个异常，以给用户
一个友好提示
            throw new UserExistException("注册的用户名已存在!!! ");
        }
        userDao.add(user);
    }
    @Override
    public List<Map<String, Object>> loginUser(String userName, String userPwd) {
        return userDao.find(userName, userPwd);
    }
}
```

DBUtils.java 代码如下。

```
package com.isoft.util;
public class DBUtils {
    static Connection conn;
    static PreparedStatement pstmt;
    static ResultSet rs;
    static public Connection getConnection() {
        try {
            Properties properties = new Properties();
            properties.load(DBUtils.class.getResourceAsStream("../../../jdbcCon-
fig.properties"));
            if(conn == null) {
                Class.forName(properties.getProperty("driverClassName"));
```

```
                    conn = DriverManager.getConnection(properties.getProperty
("url"), properties.getProperty("username"),
                                        properties.getProperty("password"));
                }
            } catch (Exception e) {
                System.out.println(" 连接数据库失败 ");
            }
            return conn;
    }
    public static void main(String[] args) {
            DBUtils.getConnection();
            // String sql = "update tb_tickerinfo set number=number+1 where
            // uname=?";
            // String sql = "insert into tb_tickerinfo(uname,number) values (?,?)";
            // int i = DBUtils.update(sql, "huangzhong", 10);
            List<Map<String, Object>> query = DBUtils.query("select sum(number) as
sum from tb_tickerinfo");
            System.out.println(query);
    }
    static public List<Map<String, Object>> query(String sql, Object... arg) {
            List<Map<String, Object>> list = new ArrayList<Map<String, Object>>();
            try {
                pstmt = conn.prepareStatement(sql);
                for (int i = 0; i < arg.length; i++) {
                    pstmt.setObject(i + 1, arg[i]);
                }
                rs = pstmt.executeQuery();
                ResultSetMetaData rsmd = rs.getMetaData();
                while (rs.next()) {
                    Map map = new HashMap();
                    for (int i = 0; i < rsmd.getColumnCount(); i++) {
                        map.put(rsmd.getColumnLabel(i + 1), rs.getObject(i
+ 1));
                    }
                    list.add(map);
                }
            } catch (Exception e) {
```

```
                e.printStackTrace();
            }
            return list;
        }
        static public int update(String sql, Object... arg) {
            try {
                pstmt = conn.prepareStatement(sql);
                for (int i = 0; i < arg.length; i++) {
                    pstmt.setObject(i + 1, arg[i]);
                }
                int i = pstmt.executeUpdate();
                return i;
            } catch (Exception e) {
                System.out.println(e.getMessage());
            }
            return 0;
        }
        static public void close() {
            try {
                if(rs != null)
                    rs.close();
                if(pstmt != null)
                    pstmt.close();
                if(conn != null)
                    conn.close();
            } catch (Exception e) {
                System.out.println(" 关闭数据库失败 ");
            } finally {
                System.out.println(" 关闭操作处理完成 ");
            }
        }
    }
```

WebUtils.java 代码如下。

```
package com.isoft.util;
public class WebUtils {
    /**
```

```
     * 将 request 对象转换成 T 对象
     */
    public static <T> T request2Bean(HttpServletRequest request,Class<T> clazz){
        try{
            T bean = clazz.newInstance();
            Enumeration<String> e = request.getParameterNames();
            while(e.hasMoreElements()){
                String name = (String) e.nextElement();
                String value = request.getParameter(name);
                BeanUtils.setProperty(bean, name, value);
            }
            return bean;
        }catch(Exception e){
            throw new RuntimeException(e);
        }
    }
    /**
     * 生成 UUID
     */
    public static String makeId(){
        return UUID.randomUUID().toString();
    }

}
```

LoginServlet.java 代码如下。

```
package com.isoft.web.controller;
@WebServlet("/loginServlet")
public class LoginServlet extends HttpServlet {
    public void doGet(HttpServletRequest request, HttpServletResponse response) throws
ServletException, IOException {
        //获取用户填写的登录用户名
        String username = request.getParameter("username");
        //获取用户填写的登录密码
        String password = request.getParameter("password");
        IUserService service = new UserServiceImpl();
        //用户登录
```

```
        List<Map<String, Object>> loginUser = service.loginUser(username, password);
        if(loginUser.size() == 0) {
            String message = String.format(
                    "对不起,用户名或密码有误!! 请重新登录! 2 秒后为您自动跳到登录页面!! <meta http-equiv='refresh' content='2;url=%s'",
                    "loginUIServlet");
            request.setAttribute("message", message);
            request.getRequestDispatcher("/message.jsp").forward(request, response);
            return;
        }
        // 登录成功后,就将用户存储到 session 中
        request.getSession().setAttribute("user", loginUser.get(0));
        String message = String.format("恭喜: %s, 登录成功! 本页将在 3 秒后跳到首页!! <meta http-equiv='refresh' content='3;url=%s'",
                loginUser.get(0).get("uname"), "index.jsp");
        request.setAttribute("message", message);
        request.getRequestDispatcher("/message.jsp").forward(request, response);
    }
    public void doPost(HttpServletRequest request, HttpServletResponse response) throws ServletException, IOException {
        doGet(request, response);
    }
}
```

LogoutServlet.java 代码如下。

```
package com.isoft.web.controller;
@WebServlet("/logoutServlet")
public class LogoutServlet extends HttpServlet {
    public void doGet(HttpServletRequest request, HttpServletResponse response) throws ServletException, IOException {
        // 移除存储在 session 中的 user 对象,实现注销功能
        request.getSession().removeAttribute("user");
        // 由于字符串中包含有单引号,在这种情况下使用 MessageFormat.format 方法拼接字符串时就会有问题
```

```
            // MessageFormat.format 方法只是把字符串中的单引号去掉，不会将内容填充到指定的占位符中
            String tempStr1 = MessageFormat.format("注销成功!! 3 秒后为您自动跳到登录页面!! <meta http-equiv='refresh' content='3;url={0}'/>",
                    "loginUIServlet");
            System.out.println(tempStr1);// 输出结果：注销成功!! 3 秒后为您自动跳到登录页面!! <meta
                    // http-equiv=refresh
                    // content=3;url={0}/>
            System.out.println("-----------------------------------------------------");
            String tempStr2 = MessageFormat
                    .format("注销成功!! 3 秒后为您自动跳到登录页面!! <meta http-equiv=\"refresh\" content=\"3;url={0}\"/>", "loginUIServlet");
            System.out.println(tempStr2);
            String message = String.format("注销成功!! 3 秒后为您自动跳到登录页面!! <meta http-equiv='refresh' content='3;url=%s'/>",
                    "loginUIServlet");
            request.setAttribute("message", message);
            request.getRequestDispatcher("/message.jsp").forward(request, response);
        }
        public void doPost(HttpServletRequest request, HttpServletResponse response) throws ServletException, IOException {
            doGet(request, response);
        }
    }
```

RegisterServlet.java 代码如下。

```
    package com.isoft.web.controller;
    @WebServlet("/registerServlet")
    public class RegisterServlet extends HttpServlet {
        public void doGet(HttpServletRequest request, HttpServletResponse response) throws ServletException, IOException {
            // 将客户端提交的表单数据封装到 RegisterFormBean 对象中
            RegisterFormBean formbean = WebUtils.request2Bean(request, Register Form Bean.class);
            // 校验用户注册填写的表单数据
            if(formbean.validate() == false) {// 如果校验失败
```

```
                //将封装了用户填写的表单数据的 formbean 对象发送回 register.jsp
页面的 form 表单中进行显示
                request.setAttribute("formbean", formbean);
                //校验失败说明是用户填写的表单数据有问题，那么就跳转回 regis-
ter.jsp
                request.getRequestDispatcher("/pages/register.jsp").forward(request,
response);
                return;
            }
            User user = new User();
            try {
                //注册字符串到日期的转换器
                ConvertUtils.register(new DateLocaleConverter(), Date.class);
                BeanUtils.copyProperties(user, formbean);// 把表单的数据填充到
javabean 中
                user.setId(WebUtils.makeId());// 设置用户的 Id 属性
                IUserService service = new UserServiceImpl();
                //调用 service 层提供的注册用户服务实现用户注册
                service.registerUser(user);
                String message = String.format("注册成功!! 3 秒后为您自动跳到登
录页面!! <meta http-equiv='refresh' content='3;url=%s'/>",
                                "loginUIServlet");
                request.setAttribute("message", message);
                request.getRequestDispatcher("/message.jsp").forward(request, response);
            } catch (UserExistException e) {
                formbean.getErrors().put("userName", "注册用户已存在!! ");
                request.setAttribute("formbean", formbean);
                request.getRequestDispatcher("/pages/register.jsp").forward(request,
response);
            } catch (Exception e) {
                e.printStackTrace(); // 在后台记录异常
                request.setAttribute("message", "对不起，注册失败!! ");
                request.getRequestDispatcher("/message.jsp").forward(request, response);
            }
        }
```

```
    public void doPost(HttpServletRequest request, HttpServletResponse response) throws
ServletException, IOException {
            doGet(request, response);
    }
}
```

CharacterEncodingFilter.java 代码如下。

```
package com.isoft.web.filter;
/**
 * 过滤器处理表单传到 Servlet 的乱码问题
 */
@WebFilter(value = "/*", initParams = { @WebInitParam(name = "encoding", value = "utf-
8")})
public class CharacterEncodingFilter implements Filter {
    // 存储系统使用的字符编码
    private String encoding = null;
    @Override
    public void init(FilterConfig filterConfig) throws ServletException {
            // encoding 在 web.xml 中指定
            this.encoding = filterConfig.getInitParameter("encoding");
    }
    @Override
    public void doFilter(ServletRequest request, ServletResponse response, FilterChain
chain)
                    throws IOException, ServletException {
            // 解决表单提交时的中文乱码问题
            request.setCharacterEncoding(encoding);
            chain.doFilter(request, response);
    }
}
```

RegisterFormBean.java 代码如下。

```
package com.isoft.web.formbean;
/**
 * 封装的用户注册表单 bean,用来接收 register.jsp 中的表单输入项的值
 * RegisterFormBean 中的属性与 register.jsp 中的表单输入项的 name 一一对应
```

```
    * RegisterFormBean 的职责除了负责接收 register.jsp 中的表单输入项的值之外，还负责校验表单输入项的值的合法性
    */
public class RegisterFormBean {
    //RegisterFormBean 中的属性与 register.jsp 中的表单输入项的 name 一一对应
    //<input type="text" name="userName"/>
    private String userName;
    //<input type="password" name="userPwd"/>
    private String userPwd;
    //<input type="password" name="confirmPwd"/>
    private String confirmPwd;
    //<input type="text" name="email"/>
    private String email;
    //<input type="text" name="birthday"/>
    private String birthday;
    /**
     * 存储校验不通过时给用户的错误提示信息
     */
    private Map<String, String> errors = new HashMap<String, String>();
    public Map<String, String> getErrors() {
            return errors;
    }
    public void setErrors(Map<String, String> errors) {
            this.errors = errors;
    }
    /*
     * validate 方法负责校验表单输入项
     * 表单输入项校验规则:
     *      private String userName; 用户名不能为空，并且是 3-8 位的字母 abcdABcd
     *      private String userPwd; 密码不能为空，并且要是 3-8 位的数字
     *      private String confirmPwd; 两次密码要一致
     *      private String email; 可以为空，不为空时是一个合法的邮箱
     *      private String birthday; 可以为空，不为空时是一个合法的日期
     */
    public boolean validate() {
            boolean isOk = true;
```

```
if(this.userName == null || this.userName.trim().equals("")) {
        isOk = false;
        errors.put("userName", " 用户名不能为空!! ");
} else {
        if(!this.userName.matches("[a-zA-Z]{3,8}")) {
                isOk = false;
                errors.put("userName", " 用户名必须是 3-8 位的字母!! ");
        }
}
if(this.userPwd == null || this.userPwd.trim().equals("")) {
        isOk = false;
        errors.put("userPwd", " 密码不能为空!! ");
} else {
        if(!this.userPwd.matches("\\d{3,8}")) {
                isOk = false;
                errors.put("userPwd", " 密码必须是 3-8 位的数字!! ");
        }
}
// private String password2; 两次密码要一致
if(this.confirmPwd != null) {
        if(!this.confirmPwd.equals(this.userPwd)) {
                isOk = false;
                errors.put("confirmPwd", " 两次密码不一致!! ");
        }
}
// private String email; 可以为空,不为空时是一个合法的邮箱
if(this.email != null && !this.email.trim().equals("")) {
        if(!this.email.matches("\\w+@\\w+(\\.\\w+)+")) {
                isOk = false;
                errors.put("email", " 邮箱不是一个合法邮箱!! ");
        }
}
// private String birthday; 可以为空,不为空时是一个合法的日期
if(this.birthday != null && !this.birthday.trim().equals("")) {
        try {
        DateLocaleConverter conver = new DateLocaleConverter();
```

```
                    conver.convert(this.birthday);
                } catch (Exception e) {
                    isOk = false;
                    errors.put("birthday", " 生日必须要是一个日期!! ");
                }
            }
            return isOk;
        }
        public String getUserName() {
            return userName;
        }
        public void setUserName(String userName) {
            this.userName = userName;
        }
        public String getUserPwd() {
            return userPwd;
        }
        public void setUserPwd(String userPwd) {
            this.userPwd = userPwd;
        }
        public String getConfirmPwd() {
            return confirmPwd;
        }
        public void setConfirmPwd(String confirmPwd) {
            this.confirmPwd = confirmPwd;
        }
        public String getEmail() {
            return email;
        }
        public void setEmail(String email) {
            this.email = email;
        }
        public String getBirthday() {
            return birthday;
        }
        public void setBirthday(String birthday) {
            this.birthday = birthday;
```

```
    }
}
```

LoginUIServlet.java 代码如下。

```
package com.isoft.web.UI;
@WebServlet("/loginUIServlet")
public class LoginUIServlet extends HttpServlet {
    public void doGet(HttpServletRequest request, HttpServletResponse response)
                throws ServletException, IOException {
            request.getRequestDispatcher("/pages/login.jsp").forward(request, response);
    }
    public void doPost(HttpServletRequest request, HttpServletResponse response)
                throws ServletException, IOException {
            doGet(request, response);
    }
}
```

RegisterUIServlet.java 代码如下。

```
package com.isoft.web.UI;
@WebServlet("/registerUIServlet")
public class RegisterUIServlet extends HttpServlet {
    public void doGet(HttpServletRequest request, HttpServletResponse response) throws
ServletException, IOException {
            request.getRequestDispatcher("/pages/register.jsp").forward(request, response);
    }
    public void doPost(HttpServletRequest request, HttpServletResponse response) throws
ServletException, IOException {
            doGet(request, response);
    }
}
```

jdbcConfig.properties 代码如下。

```
driverClassName=org.gjt.mm.mysql.Driver
url=jdbc:mysql://localhost:3306/test?useUnicode=true&characterEncoding=utf-8
username=root
password=root
```

tb_user 数据库表结构如图 13.4 所示。

栏位 索引 外键 触发器 选项 注释 SQL 预览

名	类型	长度	小数点	不是 null	
id	varchar	50	0	☑	🔑1
uname	varchar	20	0	☑	
upwd	varchar	20	0	☑	
email	varchar	50	0	☐	
birthday	date	0	0	☐	

图 13.4 tb_user 数据库表结构

运行结果如图 13.5 所示。

用户登录

用户名

admin

密 码

111111

登录 返回注册

localhost:8081/Model%20II/loginServlet

恭喜：admin,登录成功！本页将在3秒后跳到首页！！

localhost:8081/Model%20II/index.jsp

MVC Model II模式案例

欢迎您：admin 退出登录

图 13.5 运行结果

图 13.5　运行结果（续）

小结

1. Model Ⅰ 和 Model Ⅱ 体系结构用于开发 Web 应用程序。

2. 在 Model Ⅰ 体系结构中，JSP 页面单独负责开发 Web 应用程序。

3. Model Ⅰ Web 应用程序由复杂的 Web 逻辑和指向 Web 应用程序中其他页面的链接组成。

4. 建立 Model Ⅰ 体系结构是一个很费时、费力的过程。

5. Model Ⅰ 体系结构提供的安全功能也很有限。

6. 为了克服 Model Ⅰ 体系结构的缺陷，引入了 Model Ⅱ 体系结构。

7. Model Ⅱ 体系结构也称为模型 – 视图 – 控制器组件体系结构（MVC）。

8. 模型、视图和控制器是 MVC 体系结构的组件。

9. 模型对象引用应用程序中所使用的数据元素。

10. 视图对象是应用程序的图形化表示。

11. 控制器对象截取视图的请求，并将它传递给模型以执行相应的动作。

经典面试题

1. Java Web 的 MVC 模式是什么？
2. 结合 Java Web 应用开发简述优点。
3. 现在 JavaWeb 开发的主流 MVC 框架有哪些？
4. 如何设计一个 Java Web MVC 框架？
5. Java Web 项目中 MVC 模型下各层的作用是什么？

跟我上机

完成如下图所示的顺风搬家预约信息查询系统。

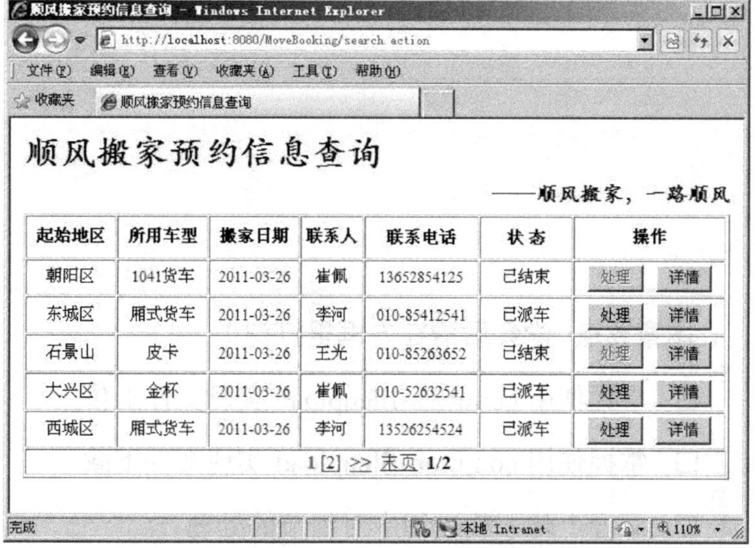

第 14 章　文件上传和下载

本章要点(学会后请在方框里打钩):

- □ 掌握使用 common-fileupload 实现文件上传
- □ 掌握使用 common-fileupload 实现文件下载
- □ 掌握上传下载过程中对细节的处理

在 Web 应用系统开发中，文件上传和下载功能是非常常用的功能，本章讲一下 JavaWeb 中的文件上传和下载功能的实现。

对于文件上传，浏览器在上传的过程中是将文件以流的形式提交到服务器端的，如果直接使用 Servlet 获取上传文件的输入流然后再解析里面的请求参数是比较麻烦的，所以一般选择采用 Apache 的开源工具 common-fileupload 这个文件上传组件。这个 common-fileupload 上传组件的 jar 包可以在 Apache 官网上面下载，也可以在 struts 的 lib 文件夹下面找到，struts 上传的功能就是基于这个实现的。因为 common-fileupload 是依赖于 common-io 这个包的，所以还需要下载这个包。

14.1 开发环境搭建

创建一个名为 FileUploadAndDownLoad 的 Maven 项目，加入配置 Apache 的 commons-fileupload 文件上传组件的 jar 包，如图 14.1 所示。

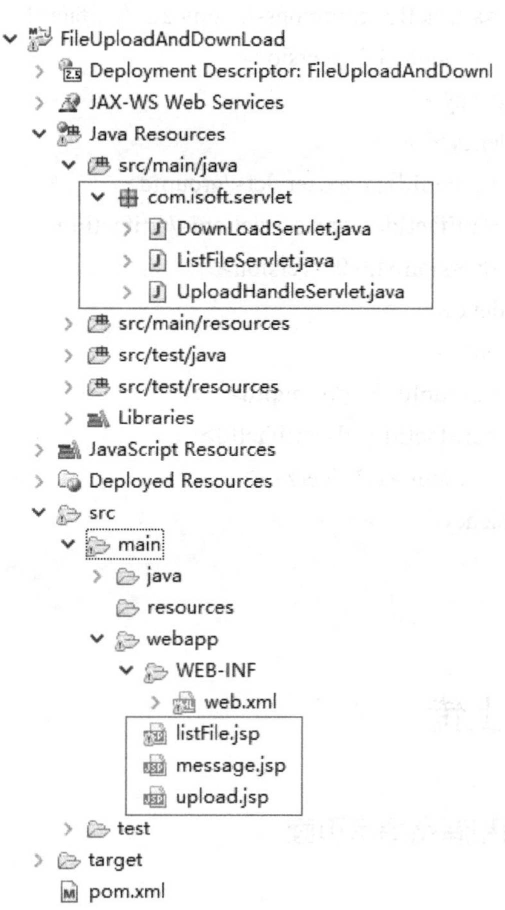

图 14.1 工程目录结构

pom.xml 代码如下。

```
<project xmlns="http://maven.apache.org/POM/4.0.0" xmlns:xsi="http://www.w3.org/2001/XMLSchema-instance"
    xsi:schemaLocation="http://maven.apache.org/POM/4.0.0 http://maven.apache.org/xsd/maven-4.0.0.xsd">
    <modelVersion>4.0.0</modelVersion>
    <groupId>com.isoft</groupId>
    <artifactId>FileUploadAndDownLoad</artifactId>
    <version>0.0.1-SNAPSHOT</version>
    <packaging>war</packaging>
    <dependencies>
        <dependency>
            <groupId>commons-fileupload</groupId>
            <artifactId>commons-fileupload</artifactId>
            <version>1.3</version>
        </dependency>
        <dependency>
            <groupId>javax.servlet</groupId>
            <artifactId>javax.servlet-api</artifactId>
            <version>4.0.0</version>
        </dependency>
        <dependency>
            <groupId>jstl</groupId>
            <artifactId>jstl</artifactId>
            <version>1.2</version>
        </dependency>
    </dependencies>
</project>
```

14.2 实现文件上传

14.2.1 文件上传页面和消息提示页面

upload.jsp 代码如下。

```
<%@ page language="java" contentType="text/html; charset=UTF-8" pageEncoding="UTF
-8"%>
<!DOCTYPE html PUBLIC "-//W3C//DTD HTML 4.01 Transitional//EN" "http://www.
w3.org/TR/html4/loose.dtd">
<html>
<head>
<meta http-equiv="Content-Type" content="text/html; charset=UTF-8">
<title> 文件上传 </title>
</head>
<body>
    <form action="${pageContext.request.contextPath}/uploadHandleServlet"
        enctype="multipart/form-data" method="post">
            上传用户： <input type="text" name="username"><br />
            上传文件 1:<input type="file" name="file1"><br />
            上传文件 2:<input type="file" name="file2"><br />
            <input type="submit" value=" 提交 ">
    </form>
</body>
</html>
```

message.jsp 的代码如下。

```
<%@ page language="java" pageEncoding="UTF-8"%>
<!DOCTYPE HTML>
<html>
<head>
<title> 消息提示 </title>
</head>
<body>${message}
</body>
</html>
```

14.2.2 处理文件上传的 Servlet

UploadHandleServlet 代码如下。

```
package com.isoft.servlet;
// 导入包略
@WebServlet("/uploadHandleServlet")
```

```
public class UploadHandleServlet extends HttpServlet {
    private static final long serialVersionUID = 1L;
    public void doGet(HttpServletRequest request, HttpServletResponse response) throws
ServletException, IOException {
        //得到上传文件的保存目录，将上传的文件存储在 WEB-INF 目录下，不允许
外界直接访问，保证上传文件的安全
        String savePath = this.getServletContext().getRealPath("/WEB-INF/upload");
        File file = new File(savePath);
        //判断上传文件的保存目录是否存在
        if(!file.exists() && !file.isDirectory()) {
            System.out.println(savePath + " 目录不存在，需要创建 ");
            file.mkdir();
        }
        //消息提示
        String message = "";
        try {
            //使用 Apache 文件上传组件处理文件上传步骤：
            //1. 创建一个 DiskFileItemFactory 工厂
            DiskFileItemFactory factory = new DiskFileItemFactory();
            //2. 创建一个文件上传解析器
            ServletFileUpload upload = new ServletFileUpload(factory);
            //解决上传文件名的中文乱码
            upload.setHeaderEncoding("UTF-8");
            //3. 判断提交上来的数据是否是上传表单的数据
            if(!ServletFileUpload.isMultipartContent(request)) {
                //按照传统方式获取数据
                return;
            }
            //4. 使用 ServletFileUpload 解析器解析上传数据，解析结果返回的
是一个 List<FileItem> 集合，每一个 FileItem 对应一个 Form 表单的输入项
            List<FileItem> list = upload.parseRequest(request);
            for(FileItem item : list) {
                //如果 FileItem 中封装的是普通输入项的数据
                if(item.isFormField()) {
                    String name = item.getFieldName();
                    //解决普通输入项的数据的中文乱码问题
                    String value = item.getString("UTF-8");
```

```
                    // value = new String(value.getBytes("iso8859-1"),
"UTF-8");
                    System.out.println(name + "=" + value);
                } else {// 如果 FileItem 中封装的是上传文件
                        //得到上传的文件名称，
                    String filename = item.getName();
                    System.out.println(filename);
                    if(filename == null || filename.trim().equals("")) {
                        continue;
                    }
                    //注意:不同的浏览器提交的文件名是不一样的,有
些浏览器提交上来的文件名是带有路径的,如:
                    //c:\a\b\1.txt,而有些只是单纯的文件名,如:1.txt
                    //处理获取到的上传文件的文件名的路径部分,只
保留文件名部分
                    filename = filename.substring(filename.lastIndexOf
("\\") + 1);
                    //获取 item 中的上传文件的输入流
                    InputStream in = item.getInputStream();
                    //创建一个文件输出流
                    FileOutputStream out = new FileOutputStream(save-
Path + "\\" + filename);
                    //创建一个缓冲区
                    byte buffer[] = new byte[1024];
                    //判断输入流中的数据是否已经读完的标识
                    int len = 0;
                    //循环将输入流读入缓冲区中,(len=in.read(buf-
fer))>0 就表示 in 里面还有数据
                    while((len = in.read(buffer)) > 0) {
                        out.write(buffer, 0, len);
                    }
                    in.close();
                    out.close();
                    item.delete();
                    message = " 文件上传成功！";
                }
            }
```

```
            } catch (Exception e) {
                    message = " 文件上传失败！";
                    e.printStackTrace();
            }
            request.setAttribute("message", message);
            request.getRequestDispatcher("/message.jsp").forward(request, response);
    }

    public void doPost(HttpServletRequest request, HttpServletResponse response) throws
ServletException, IOException {
            doGet(request, response);
    }
}
```

运行结果如图 14.2 所示。

上传用户：	融创软通IT学院
上传文件1：	E:\E盘\工具\2336_image(浏览...
上传文件2：	E:\E盘\工具\融创软通LOC 浏览...
提交	

图 14.2 运行结果

文件上传成功之后，上传的文件保存在了 WEB-INF 目录下的 upload 目录中，如图 14.3 所示。

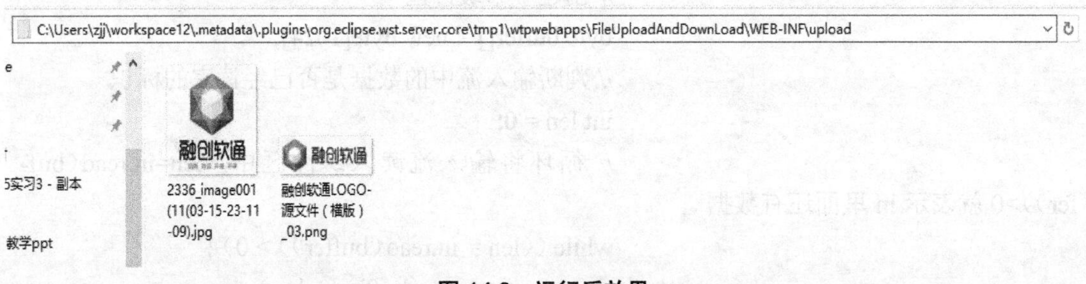

图 14.3 运行后效果

14.2.3 文件上传的细节

上述代码虽然可以成功将文件上传到服务器上面的指定目录中，但是文件上传功能有许多需要注意的小细节问题，以下列出的几点需要特别注意。

（1）为保证服务器安全，上传文件应该存储在外界无法直接访问的目录下，比如存储在 WEB-INF 目录下。

（2）为防止文件覆盖的现象发生，要为上传文件产生一个唯一的文件名。

(3)为防止一个目录下面出现太多文件,要使用 hash 算法打散存储。

(4)要限制上传文件的最大值。

(5)要限制上传文件的类型,在收到上传文件名时,判断后缀名是否合法。

针对上述五点细节问题,下面改进 UploadHandleServlet,改进后的代码如下。

```
//导入包略
@WebServlet("/uploadHandleServlet")
public class UploadHandleServlet extends HttpServlet {
    private static final long serialVersionUID = 1L;
    public void doGet(HttpServletRequest request, HttpServletResponse response) throws
ServletException, IOException {
            //得到上传文件的保存目录,将上传的文件存储于 WEB-INF 目录下,不允许
外界直接访问,保证上传文件的安全
            String savePath = this.getServletContext().getRealPath("/WEB-INF/upload");
            //上传时生成的临时文件保存目录
            String tempPath = this.getServletContext().getRealPath("/WEB-INF/temp");
            File tmpFile = new File(tempPath);
            if(!tmpFile.exists()) {
                //创建临时目录
                tmpFile.mkdir();
            }
            //消息提示
            String message = "";
            try {
                //使用 Apache 文件上传组件处理文件上传步骤:
                //1. 创建一个 DiskFileItemFactory 工厂
                DiskFileItemFactory factory = new DiskFileItemFactory();
                //设置工厂的缓冲区的大小,当上传的文件大小超过缓冲区的大小
时,就会生成一个临时文件存储到指定的临时目录当中
                factory.setSizeThreshold(1024 * 100);// 设置缓冲区的大小为
100 KB,如果不指定,那么缓冲区的大小默认是 10 KB
                //设置上传时生成的临时文件的保存目录
                factory.setRepository(tmpFile);
                //2. 创建一个文件上传解析器
                ServletFileUpload upload = new ServletFileUpload(factory);
                //监听文件上传进度
                upload.setProgressListener(new ProgressListener() {
```

```
                    public void update(long pBytesRead, long pContentLength, int arg2) {
                        System.out.println(" 文件大小为:" + pContentLength + ", 当前已处理:" + pBytesRead);
                    }
                });
                //解决上传文件名的中文乱码
                upload.setHeaderEncoding("UTF-8");
                //3. 判断提交上来的数据是否是上传表单的数据
                if(!ServletFileUpload.isMultipartContent(request)) {
                    //按照传统方式获取数据
                    return;
                }
                //设置上传单个文件的大小的最大值,目前是设置为 1024*1024 字节,也就是 1 MB
                upload.setFileSizeMax(1024 * 1024);
                //设置上传文件总量的最大值,最大值 = 同时上传的多个文件的大小的最大值的和,目前设置为 10 MB
                upload.setSizeMax(1024 * 1024 * 10);
                //4. 使用 ServletFileUpload 解析器解析上传数据,解析结果返回的是一个 List<FileItem> 集合,每一个 FileItem 对应一个 Form 表单的输入项
                List<FileItem> list = upload.parseRequest(request);
                for(FileItem item : list) {
                    //如果 FileItem 中封装的是普通输入项的数据
                    if(item.isFormField()) {
                        String name = item.getFieldName();
                        //解决普通输入项的数据的中文乱码问题
                        String value = item.getString("UTF-8");
                        // value = new String(value.getBytes("iso8859-1"), "UTF-8");
                        System.out.println(name + "=" + value);
                    } else {// 如果 FileItem 中封装的是上传文件
                        //得到上传的文件名称,
                        String filename = item.getName();
                        System.out.println(filename);
```

```
if(filename == null || filename.trim().equals("")) {
    continue;
}
// 注意：不同的浏览器提交的文件名是不一样的，有些浏览器提交上来的文件名是带有路径的，如：
// c:\a\b\1.txt，而有些只是单纯的文件名，如：1.txt
// 处理获取到的上传文件的文件名的路径部分，只保留文件名部分
filename = filename.substring(filename.lastIndexOf("\\") + 1);
// 得到上传文件的扩展名
String fileExtName = filename.substring(filename.lastIndexOf(".") + 1);
// 如果需要限制上传的文件类型，那么可以通过文件的扩展名来判断上传的文件类型是否合法
System.out.println("上传的文件的扩展名是：" + fileExtName);
// 获取 item 中的上传文件的输入流
InputStream in = item.getInputStream();
// 得到文件保存的名称
String saveFilename = makeFileName(filename);
// 得到文件的保存目录
String realSavePath = makePath(saveFilename, savePath);
// 创建一个文件输出流
FileOutputStream out = new FileOutputStream(realSavePath + "\\" + saveFilename);
// 创建一个缓冲区
byte buffer[] = new byte[1024];
// 判断输入流中的数据是否已经读完的标识
int len = 0;
// 循环将输入流读入到缓冲区当中，(len=in.read(buffer))>0 就表示 in 里面还有数据
while ((len = in.read(buffer)) > 0) {
    // 使用 FileOutputStream 输出流将缓冲区的数据写入到指定的目录(savePath + "\\"
```

```
                                        //+filename)当中
                                        out.write(buffer, 0, len);
                                }
                                in.close();
                                out.close();
                                message = "文件上传成功！";
                        }
                }
        } catch (FileUploadBase.FileSizeLimitExceededException e) {
                e.printStackTrace();
                request.setAttribute("message", "单个文件超出最大值!!! ");
                request.getRequestDispatcher("/message.jsp").forward(request, response);
                return;
        } catch (FileUploadBase.SizeLimitExceededException e) {
                e.printStackTrace();
                request.setAttribute("message", "上传文件的总的大小超出限制的最
大值!!! ");
                request.getRequestDispatcher("/message.jsp").forward(request, response);
                return;
        } catch (Exception e) {
                message = "文件上传失败！";
                e.printStackTrace();
        }
        request.setAttribute("message", message);
        request.getRequestDispatcher("/message.jsp").forward(request, response);
    }
    private String makeFileName(String filename) { // 2.jpg
        //为防止文件覆盖的现象发生，要为上传文件产生一个唯一的文件名
        return UUID.randomUUID().toString() + "_" + filename;
    }
    /**
     * 为防止一个目录下面出现太多文件，要使用 hash 算法打散存储
     */
    private String makePath(String filename, String savePath) {
        //得到文件名的 hashCode 的值，得到的就是 filename 这个字符串对象在内
存中的地址
```

```
            int hashcode = filename.hashCode();
            int dir1 = hashcode & 0xf; // 0-15
            int dir2 = (hashcode & 0xf0) >> 4; // 0-15
            // 构造新的保存目录
            String dir = savePath + "\\" + dir1 + "\\" + dir2; // upload\2\3
                                                        // upload\3\5
            // File 既可以代表文件也可以代表目录
            File file = new File(dir); // 如果目录不存在
            if(!file.exists()) { // 创建目录
                    file.mkdirs();
            }
            return dir;
        }
        public void doPost(HttpServletRequest request, HttpServletResponse response) throws ServletException, IOException {
            doGet(request, response);
        }
    }
```

控制台输出结果如图 14.4 所示。

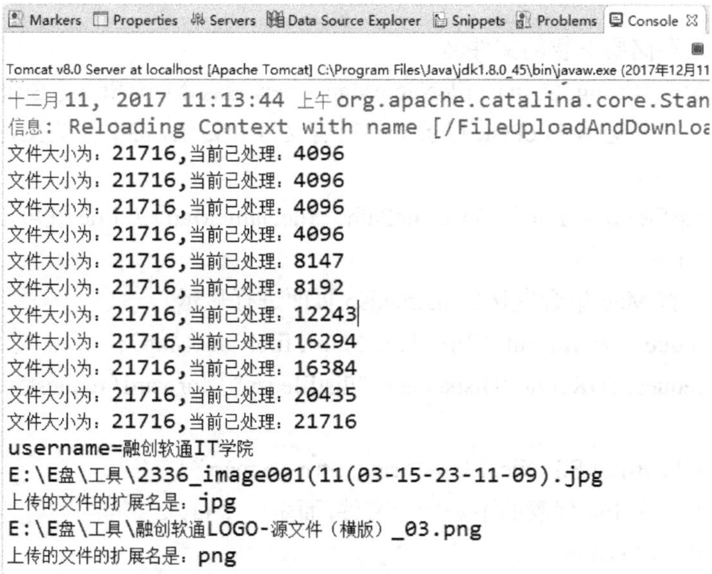

```
Markers  Properties  Servers  Data Source Explorer  Snippets  Problems  Console
Tomcat v8.0 Server at localhost [Apache Tomcat] C:\Program Files\Java\jdk1.8.0_45\bin\javaw.exe (2017年12月11
十二月 11, 2017 11:13:44 上午 org.apache.catalina.core.Stan
信息: Reloading Context with name [/FileUploadAndDownLoa
文件大小为：21716,当前已处理：4096
文件大小为：21716,当前已处理：4096
文件大小为：21716,当前已处理：4096
文件大小为：21716,当前已处理：4096
文件大小为：21716,当前已处理：8147
文件大小为：21716,当前已处理：8192
文件大小为：21716,当前已处理：12243
文件大小为：21716,当前已处理：16294
文件大小为：21716,当前已处理：16384
文件大小为：21716,当前已处理：20435
文件大小为：21716,当前已处理：21716
username=融创软通IT学院
E:\E盘\工具\2336_image001(11(03-15-23-11-09).jpg
上传的文件的扩展名是：jpg
E:\E盘\工具\融创软通LOGO-源文件（横版）_03.png
上传的文件的扩展名是：png
```

图 14.4 控制台输出结果

改进之后的文件上传功能就做得比较完善了。

14.3 实现文件下载

14.3.1 列出提供下载的文件资源

要将 Web 应用系统中的文件资源提供给用户进行下载，首先要有一个页面列出上传文件目录下的所有文件，当用户单击文件下载超链接时就进行下载操作，编写一个 ListFileServlet，用于列出 Web 应用系统中所有下载文件。

ListFileServlet. java 的源代码如下。

```
package com.isoft.servlet;
// 导包略
@WebServlet("/listFileServlet")
public class ListFileServlet extends HttpServlet {
    private static final long serialVersionUID = 1L;

    public void doGet(HttpServletRequest request, HttpServletResponse response) throws ServletException, IOException {
            // 获取上传文件的目录
            String uploadFilePath = this.getServletContext().getRealPath("/WEB-INF/upload");
            // 存储要下载的文件名
            Map<String, String> fileNameMap = new HashMap<String, String>();
            // 递归遍历 FilePath 目录下的所有文件和目录，将文件的文件名存储到 Map 集合中
            listfile(new File(uploadFilePath), fileNameMap);// File 既可以代表一个文件也可以代表一个目录
            // 将 Map 集合发送到 listfile.jsp 页面进行显示
            request.setAttribute("fileNameMap", fileNameMap);
            request.getRequestDispatcher("/listfile.jsp").forward(request, response);
    }
    public void listfile(File file, Map<String, String> map) {
            // 如果 file 代表的不是一个文件，而是一个目录
            if(!file.isFile()) {
                // 列出该目录下的所有文件和目录
                File files[] = file.listFiles();
                // 遍历 files[] 数组
```

```
                for (File f : files) {
                        // 递归
                        listfile(f, map);
                }
            } else {
                /**
                 * 处理文件名,上传后的文件是以 uuid_ 文件名的形式去重新命名
的,去除文件名的 uuid_ 部分
                 * file.getName().indexOf("_")检索字符串中第一次出现"_"字符
的位置,如果文件名类似于:9349249849-88343-8344_阿_凡_达.avi
                 * 那么 file.getName().substring(file.getName().indexOf("_")+1)
处理之后就可以得到阿_凡_达.avi 部分
                 */
                String realName = file.getName().substring(file.getName().indexOf
("_") + 1);
                // file.getName()得到的是文件的原始名称,这个名称是唯一的,因
此可以作为 key,realName 是处理过后的名称,有可能会重复
                map.put(file.getName(), realName);
            }
        }
        public void doPost(HttpServletRequest request, HttpServletResponse response) throws
ServletException, IOException {
            doGet(request, response);
        }
    }
```

ListFileServlet 中 listfile 方法是用来列出目录下的所有文件的，listfile 方法内部用到了递归，在实际开发当中，我们肯定会在数据库创建一张表，里面会存储上传的文件名以及文件的具体存放目录，通过查询表就可以知道文件的具体存放目录是不需要用到递归操作的，这个例子是因为没有使用数据库存储上传的文件名和文件的具体存放位置，而上传文件的存放位置又使用了散列算法打散存放，所以需要用到递归。在递归时，将获取到的文件名存放到从外面传递到 listfile 方法里面的 Map 集合当中，这样就可以保证所有的文件都存放在同一个 Map 集合当中。

下载文件的 listfile.jsp 代码如下。

```
<%@ page language="java" import="java.util.*" pageEncoding="UTF-8"%>
<%@taglib prefix="c" uri="http://java.sun.com/jsp/jstl/core"%>
<!DOCTYPE HTML>
```

```
<html>
<head>
<title> 下载文件显示页面 </title>
</head>
<body>
    <!-- 遍历 Map 集合 -->
    <c:forEach var="me" items="${fileNameMap}">
            <c:url value="downLoadServlet" var="downurl">
                    <c:param name="filename" value="${me.key}"></c:param>
            </c:url>
     ${me.value}<a href="${downurl}"> 下载 </a>
            <br />
    </c:forEach>
</body>
</html>
```

访问 http://localhost:8081/FileUploadAndDownLoad/listFileServlet，就可以在 listfile.jsp 页面中显示提供给用户下载的文件资源，如图 14.5 所示。

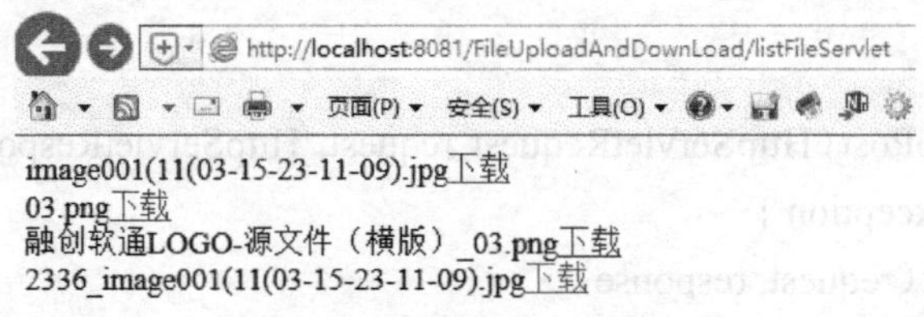

图 14.5　可下载文件列表

14.3.2　实现文件下载

编写一个用于处理文件下载的 Servlet，DownLoadServlet 的代码如下。

```
package com.isoft.servlet;
@WebServlet("/downLoadServlet")
public class DownLoadServlet extends HttpServlet {
    private static final long serialVersionUID = 1L;
    public void doGet(HttpServletRequest request, HttpServletResponse response) throws
ServletException, IOException {
            String fileName = request.getParameter("filename"); // 23239283-92489- 阿凡
达 .avi
            fileName = new String(fileName.getBytes("iso8859-1"), "UTF-8");
```

```
            String fileSaveRootPath = this.getServletContext().getRealPath("/WEB-INF/
upload");
            String path = findFileSavePathByFileName(fileName, fileSaveRootPath);
            File file = new File(path + "\\" + fileName);
            if(!file.exists()) {
                request.setAttribute("message", "您要下载的资源已被删除!! ");
                request.getRequestDispatcher("/message.jsp").forward(request, re-
sponse);
                return;
            }
            String realname = fileName.substring(fileName.indexOf("_") + 1);
            response.setHeader("content-disposition","attachment;filename="+ URLEncod-
er.encode(realname, "UTF-8"));
            FileInputStream in = new FileInputStream(path + "\\" + fileName);
            OutputStream out = response.getOutputStream();
            byte buffer[] = new byte[1024];
            int len = 0;
            while((len = in.read(buffer)) > 0) {
                out.write(buffer, 0, len);
            }
            in.close();
            out.close();
    }
    public String findFileSavePathByFileName(String filename, String saveRootPath) {
            int hashcode = filename.hashCode();
            int dir1 = hashcode & 0xf; // 0--15
            int dir2 = (hashcode & 0xf0) >> 4; // 0-15
            String dir = saveRootPath + "\\" + dir1 + "\\" + dir2; // upload\2\3
            File file = new File(dir);
            if(!file.exists()) {
                file.mkdirs();
            }
            return dir;
    }
    public void doPost(HttpServletRequest request, HttpServletResponse response) throws
ServletException, IOException {
```

```
            doGet(request, response);
        }
    }
```

单击“下载”超链接，将请求提交到 DownLoadServlet 处理就可以实现文件的下载了，运行结果如图 14.6 所示。

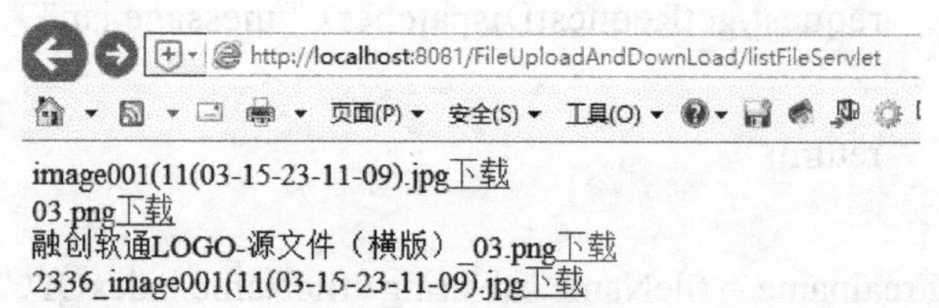

图 14.6　运行结果

从运行结果可以看到，文件下载功能已经可以正常使用了。

小结

本章主要讲述了使用 common-fileupload 组件完成对文件的上传和下载，在对文件上传和下载时要处理很多细节，需要在学习过程中多参考网上的一些案例，以便能够从容应对。

经典面试题

1. 如何用 commons-fileupload 实现文件上传？
2. 如何通过 JSP 组件 commons-fileupload 实现文件上传功能？
3.commons-fileupload 可以实现异步上传吗？
4. 如何实现文件下载功能？
5. 文件上传时如何用 commons-fileupload 限制上传文件的大小？
6. 文件上传时如何用 commons-fileupload 限制上传文件的类型？
7. 如何进行多文件上传？

跟我上机

完成下图所示功能，并将信息注册到数据库中。

图片文件上传

说明：本程序支持所有类型文件的上传；但目前只支持图片文件的查看，支持类型：bmp、gif、jpg、jpeg、png。

文件 1: [ ] 浏览...
文件 2: [ ] 浏览...
文件 3: [ ] 浏览...

上传 重选